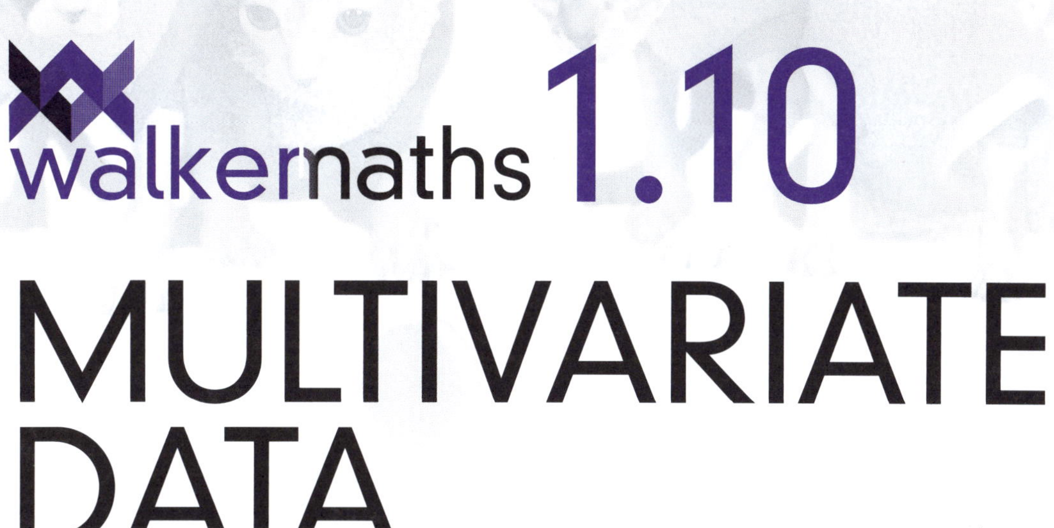

MULTIVARIATE DATA

NCEA Level 1 Internal

Charlotte Walker and Victoria Walker

Walker Maths 1.10 Multivariate Data
1st Edition
Charlotte Walker
Victoria Walker

Editor: Eva Chan
Designer: Cheryl Smith, Macarn Design
Production controller: Siew Han Ong

Any URLs contained in this publication were checked for currency during the production process. Note, however, that the publisher cannot vouch for the ongoing currency of URLs.

Acknowledgements
Cover photo courtesy of Shutterstock.

We wish to thank the Boards of Trustees of Darfield and Riccarton High Schools for allowing us to use materials and ideas developed while teaching, Our thanks also go to all past and present colleagues who have generously shared their expertise and ideas,

For product information and technology assistance,
in Australia call **1300 790 853**;
in New Zealand call **0800 449 725**

For permission to use material from this text or product, please email **aust.permissions@cengage.com**

National Library of New Zealand Cataloguing-in-Publication Data
A catalogue record for this book is available from the National Library of New Zealand.

978 0 17 037041 7

Cengage Learning Australia
Level 7, 80 Dorcas Street
South Melbourne, Victoria Australia 3205

Cengage Learning New Zealand
Unit 4B Rosedale Office Park
331 Rosedale Road, Albany, North Shore 0632, NZ

For learning solutions, visit **cengage.co.nz**

Printed in China by 1010 Printing International Limited
11 12 13 14 15 16 26 25 24 23 22

CONTENTS

ISBN: 9780170370417

Glossary

Make your own glossary of key terms:

Term	Definition	Picture/Example
Variable		
Multivariate		
Population		
Sample		
Census		
Spread		
Shape		
Unusual point		
Skew		
Maximum		

ISBN: 9780170370417

Term	Definition	Picture/Example
Upper quartile (UQ)		
Median		
Lower quartile (LQ)		
Minimum		
Inter-quartile range (IQR)		
Box plot		
Bias		
Inference		
Features		
Difference Between Means (DBM)		
Overall Visual Spread (OVS)		

ISBN: 9780170370417

Introduction

This standard will require you to investigate a multivariate data set and compare two or more subgroups. More specifically you will be required to:

- investigate data that has been collected from a survey
- pose a statistical question
- select and use appropriate display(s)
- give summary statistics
- discuss features of distributions comparatively
- make a call about whether there is a difference between two sub-groups
- communicate findings in a conclusion.

What is multivariate data?

- This is where you have several different variables for each individual in a sample or population.
- You select **one** continuous variable and compare **two** different sub-groups, e.g. you could compare weights of males and females.

Statistical enquiry cycle

Your report should connect back to all aspects of the statistical enquiry cycle:

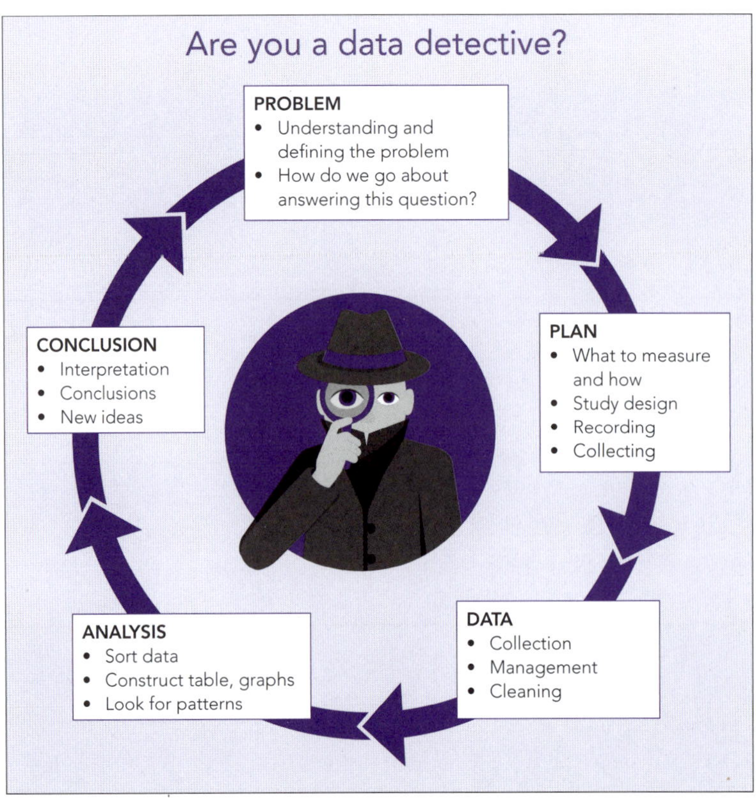

 ISBN: 9780170370417

Statistics

Measures of centre

These give a measure of where the **middle** of a distribution lies.

Name	Calculation	Advantages	Disadvantages
Mean	$\frac{\text{the sum of all the data values}}{\text{the number of data values}}$	• Easy to calculate	• Distorted by very big or small values
Median	middle value	• Usually a good measure of centre	• Data must be put in order first
Mode	value that occurs most frequently	• Very easy to find	• Very unreliable as a measure of centre • Often there are two or none

Example: 45, 27, 39, 34, 97, 33, 46, 28, 22, 23, 50, 47, 48, 33, 37, 46, 19, 22

Put the values in order: 19, 22, 22, 23, 27, 28, 33, 33, 34, 37, 39, 45, 46, 46, 47, 48, 50, 97

Median $= \frac{34 + 37}{2}$

If there are more than two 'modal values', then there is no mode.

Mean $= \frac{696}{18} = 38.67$ Median $= \frac{34 + 37}{2} = 35.5$ Modal values = 22, 33 and 46

The best measure of centre is the median. Apart from the 97, all the data is fairly evenly spread between 19 and 50, and the midpoint of these is about 35, which is close to the median. The mean has been pulled up by just one value: the 97.

Calculate the mean, median and mode for the following sets of data, and state which you think is the best measure of centre and why.

1 **1, 2, 5, 2, 4, 4, 3, 1, 1, 3**

____

Mean = ________ Median = ________ Mode = ________

The best is/are the ________________ because ________________

____

2 **18, 2, 19, 17, 14, 20, 16, 21, 20, 23**

____

Mean = ________ Median = ________ Mode = ________

The best is/are the ________________ because ________________

____

ISBN: 9780170370417

Measures of spread

These give a measure of how widely **spread** the data is.

Quartiles
The upper quartile (UQ) is the middle value of the top half of the data.
The lower quartile (LQ) is the middle value of the bottom half of the data.

Interquartile Range (IQR) = UQ – LQ: This is a very good measure of spread because extreme values do not affect it.

Range = Maximum – Minimum: This is not a good measure of spread because it is affected by very big and very small values.

Example 1:

4, 9, 5, 8, 6, 0, 3, 8, 7, 1, 7, 5, 8, 2, 2, 6, 7, 6, 7, 0

Order the data:

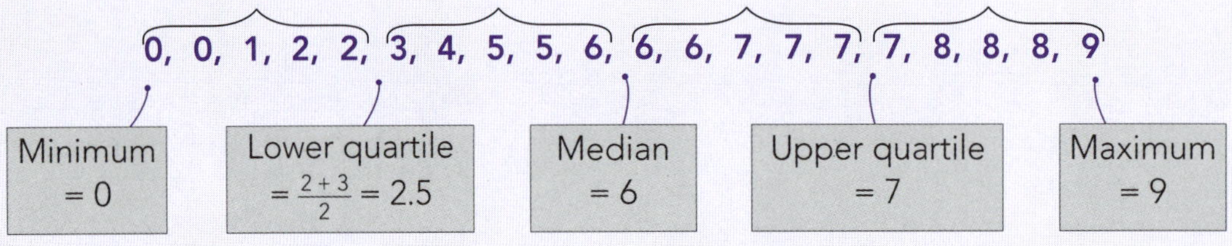

Range = 9 – 0
= **9**

IQR = 7 – 2.5
= **4.5**

Example 2:

4, 9, 5, 8, 6, 0, 3, 8, 7, 7, 5, 8, 2, 2, 3

Order the data:

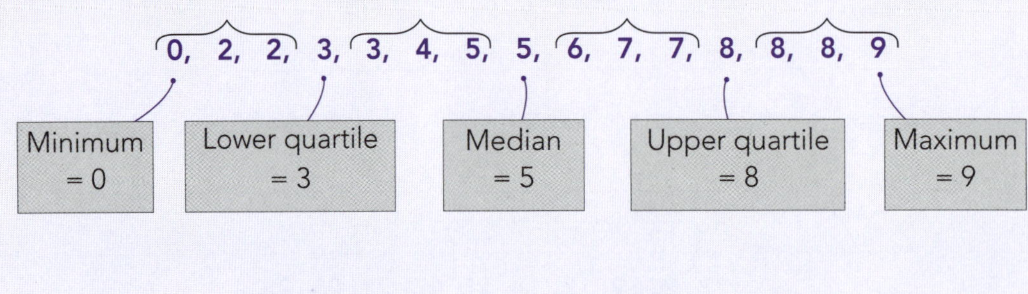

Range = 9 – 0
= **9**

IQR = 8 – 3
= **5**

ISBN: 9780170370417

On your graphics calculator

→ Menu

→ Stat (input your data)

Hint: The data does not need to be in order

→ Calc

→ 1 Var

Scroll down

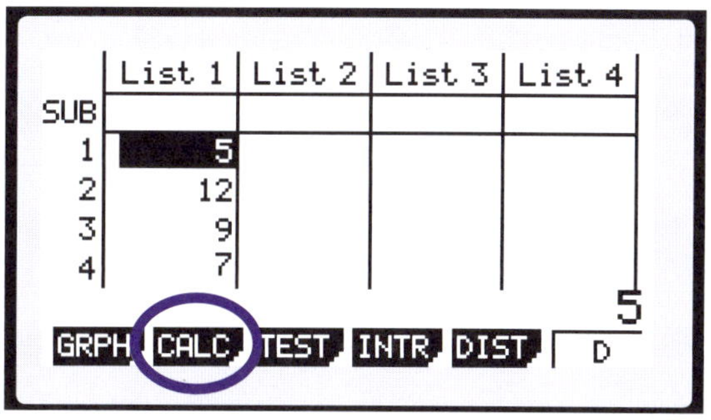

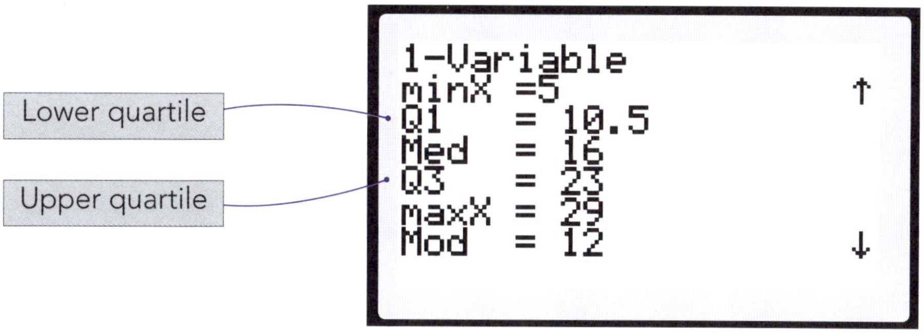

Lower quartile

Upper quartile

Calculate the statistics for the following data sets.

1 **1, 2, 2, 3, 4, 5, 5, 6, 7, 7, 8, 10, 11, 12, 12**

Minimum ______ Lower quartile ______ Median ______ Upper quartile ______

Maximum ______ Range ______ Inter-quartile range ______

2 **15, 17, 17, 17, 18, 19, 19, 19, 21, 21, 22, 23, 24, 25, 25, 25, 25, 26, 26**

Minimum ______ Lower quartile ______ Median ______ Upper quartile ______

Maximum ______ Range ______ Inter-quartile range ______

3 **0, 0, 1, 2, 2, 2, 4, 5, 5, 5, 6, 6, 7, 7, 8, 9, 9**

Minimum ______ Lower quartile ______ Median ______ Upper quartile ______

Maximum ______ Range ______ Inter-quartile range ______

ISBN: 9780170370417

4 28, 28, 29, 30, 31, 31, 33, 34, 34, 35, 36, 38, 42, 43, 44, 44, 45, 46

Minimum ______ Lower quartile ______ Median ______ Upper quartile ______

Maximum ______ Range ______ Inter-quartile range ______

5 1.5, 1.6, 1.6, 1.6, 1.8, 1.8, 1.8, 1.9, 1.9, 2.0, 2.2, 2.3, 2.4, 2.5

Minimum ______ Lower quartile ______ Median ______ Upper quartile ______

Maximum ______ Range ______ Inter-quartile range ______

6 5, 12, 15, 23, 24, 34, 39, 43, 44, 46, 49, 53, 55, 56, 59, 89, 92, 99

Minimum ______ Lower quartile ______ Median ______ Upper quartile ______

Maximum ______ Range ______ Inter-quartile range ______

7 12, 3, 24, 9, 4, 1, 12, 14, 11, 3, 5, 6, 4, 7, 9, 11, 15, 2, 5

Minimum ______ Lower quartile ______ Median ______ Upper quartile

Maximum ______ Range ______ Inter-quartile range ______

8 26, 24, 21, 28, 24, 21, 22, 20, 25, 19, 18, 24, 25, 24, 26, 27

Minimum ______ Lower quartile ______ Median ______ Upper quartile ______

Maximum ______ Range ______ Inter-quartile range ______

ISBN: 9780170370417

9 **3, 2.5, 1, 0, 7, 2.5, 4, 0, 3.5, 4.5, 2.5, 1.5, 9, 5, 7, 1, 2, 3.5, 4.5**

Minimum ______ Lower quartile ______ Median ______ Upper quartile ______

Maximum ______ Range ______ Inter-quartile range ______

10 **105, 87, 54, 26, 84, 57, 56, 23, 28, 42, 112, 23, 26, 48, 36, 33, 85**

Minimum ______ Lower quartile ______ Median ______ Upper quartile ______

Maximum ______ Range ______ Inter-quartile range ______

11 **41, 68, 55, 93, 76, 21, 44, 51, 56, 71, 69, 39, 50, 59, 62, 87, 61, 55, 77, 59**

Minimum ______ Lower quartile ______ Median ______ Upper quartile ______

Maximum ______ Range ______ Inter-quartile range ______

12 **0.4, 1.8, 2.6, 2.6, 0.8, 0.2, 0.5, 0.7, 1.1, 2.3, 2.5, 2.3, 2.6, 0.9, 0.6, 0.2, 1.3, 1.7, 0.8, 1.8, 2.0**

Minimum ______ Lower quartile ______ Median ______ Upper quartile ______

Maximum ______ Range ______ Inter-quartile range ______

13 **0.7, 4.9, 1.45, 1.65, 3.25, 0.55, 4.1, 6.5, 0.95, 2.3, 1.05, 1.1, 5.15, 4.95, 1.25, 2, 3.85, 5.2, 4.75, 2.95**

Minimum ______ Lower quartile ______ Median ______ Upper quartile ______

Maximum ______ Range ______ Inter-quartile range ______

ISBN: 9780170370417

Display of data

Dot plots

A dot plot provides a visual representation of the data.

Example: 2, 3, 3, 5, 5, 5, 6, 7, 7, 8, 8, 8, 8, 9, 9, 9, 10, 10, 12, 13, 13, 14, 15, 18

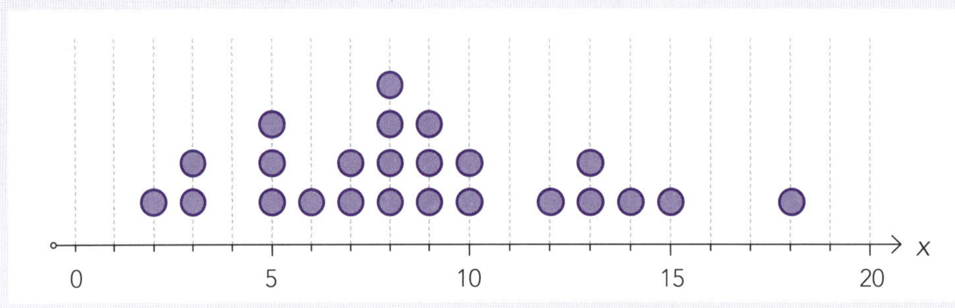

Create dot plots for these data sets.

1 2, 4, 4, 6, 7, 7, 8, 8, 9, 9, 9, 10, 10, 11, 11, 11, 12, 12, 13, 19

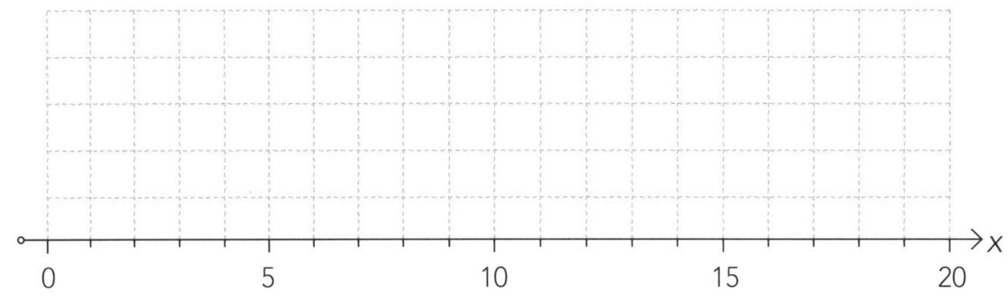

2 4, 5, 8, 15, 13, 11, 6, 12, 7, 8, 7, 13, 17, 12, 9, 8, 12, 13, 8, 13

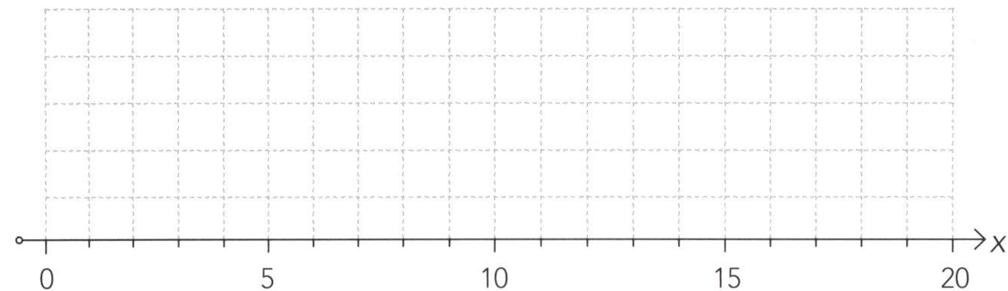

3 0, 5, 2, 8, 1, 6, 2, 2, 4, 0, 1, 9, 11, 3, 4, 2, 1, 1, 3, 3

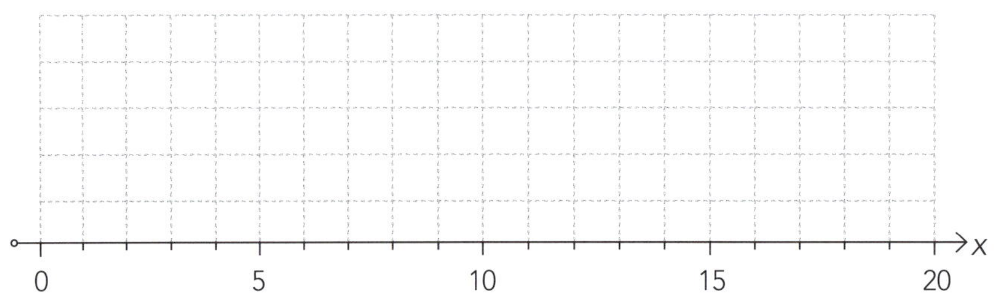

ISBN: 9780170370417

Box plots

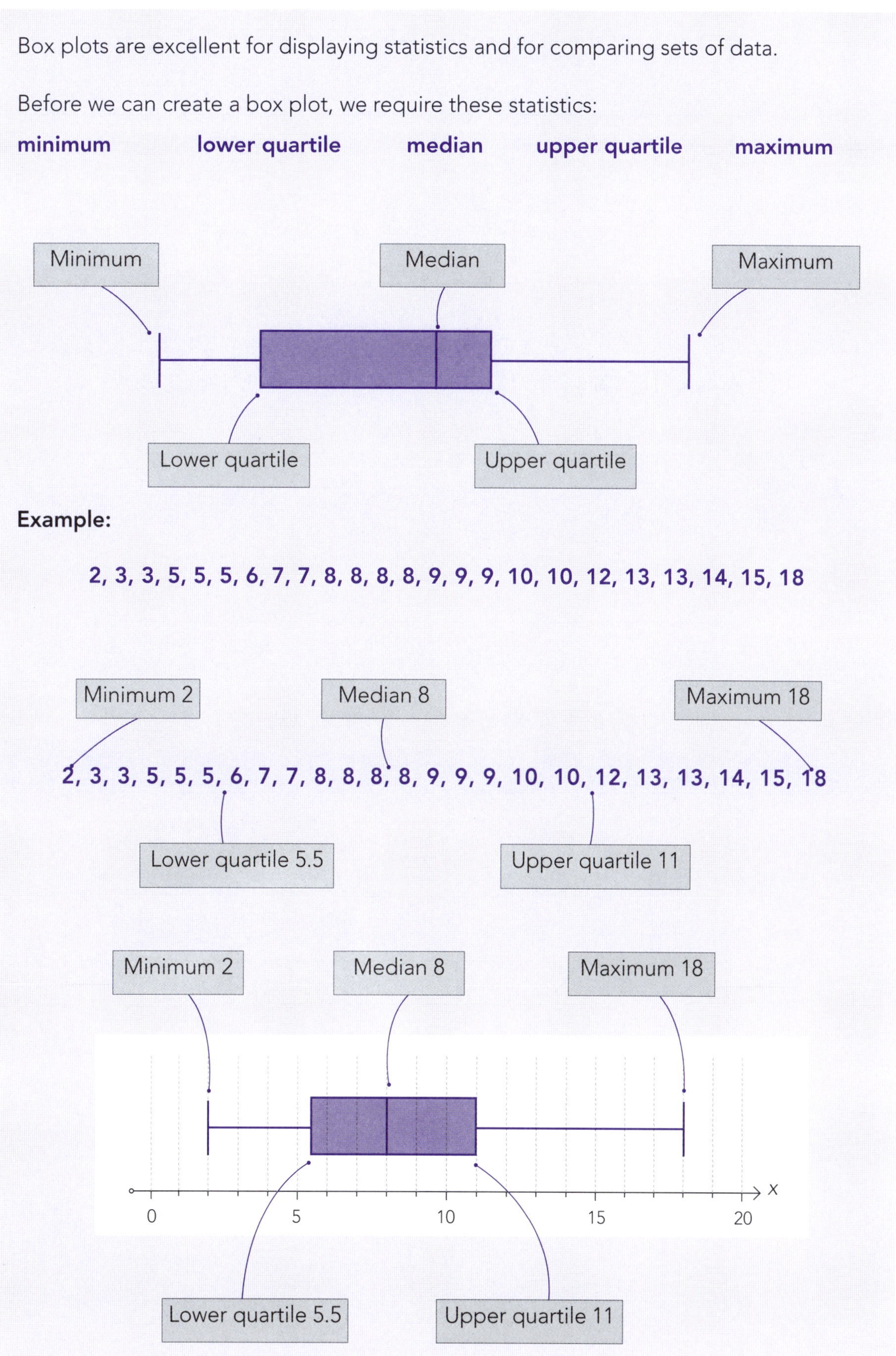

Box plots are excellent for displaying statistics and for comparing sets of data.

Before we can create a box plot, we require these statistics:

minimum **lower quartile** **median** **upper quartile** **maximum**

Example:

2, 3, 3, 5, 5, 5, 6, 7, 7, 8, 8, 8, 8, 9, 9, 9, 10, 10, 12, 13, 13, 14, 15, 18

Draw box plots for the following sets of data.

1 2, 4, 4, 6, 7, 7, 8, 8, 9, 9, 9, 10, 10, 11, 11, 11, 12, 12, 13, 19

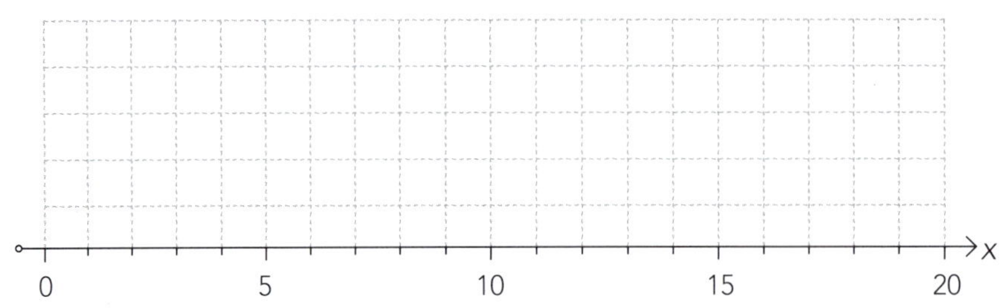

2 4, 5, 8, 15, 13, 11, 6, 12, 7, 8, 7, 13, 17, 12, 9, 8, 12, 13, 8, 13

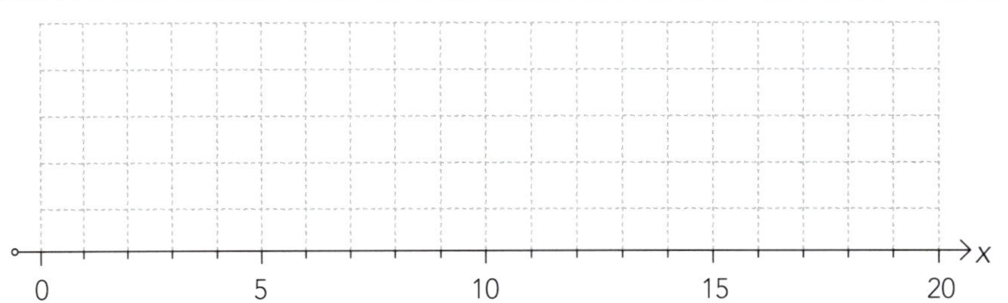

3 0, 5, 2, 8, 1, 6, 2, 2, 4, 0, 1, 9, 11, 3, 4, 2, 1, 1, 3, 3

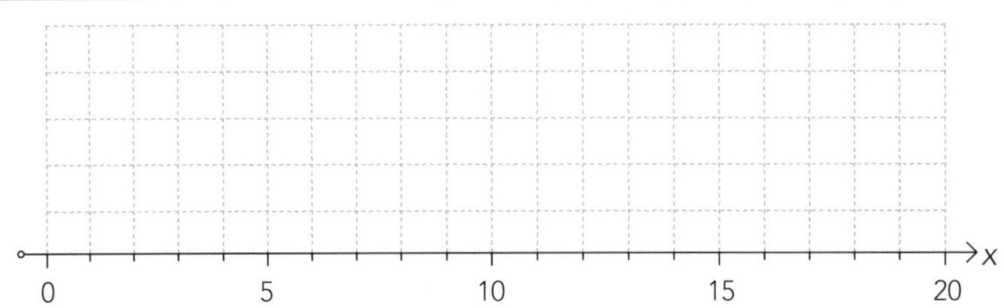

4 6, 5, 7, 8, 1, 6, 2, 12, 4, 8, 4, 9, 11, 13, 4, 2, 7, 10, 3, 3

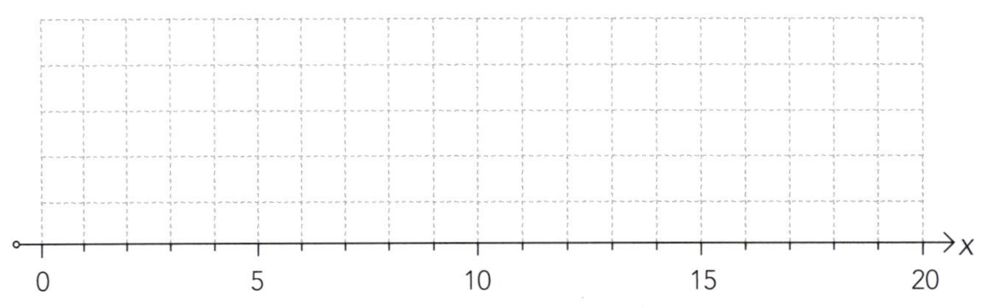

ISBN: 9780170370417

Putting it together

It is useful to draw both a box plot and a dot plot on the same scale.

Example one:

1, 2, 2, 2, 3, 4, 4, 5, 5, 5, 5, 5, 6, 6, 6, 7, 8, 8, 9, 9

Minimum 1 Lower quartile 3.5 Median 5 Upper quartile 6.5 Maximum 9

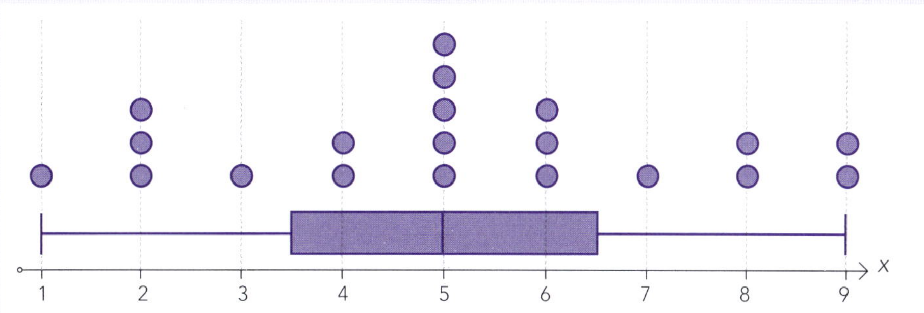

It is also useful to plot two data sets on one set of axes in order to compare distributions.

Example two:
A group of Year 11 students got their results back from a recent test.

Science:

2, 5, 7, 9, 8, 12, 14, 6, 13, 15, 7, 9, 6, 5, 7, 13, 6, 12

Minimum 2 Lower quartile 6 Median 7.5 Upper quartile 12 Maximum 15

English:

3, 12, 14, 5, 14, 13, 12, 14, 15, 3, 3, 7, 8, 9, 6, 10, 8, 7, 12, 11, 9, 11, 15, 14, 12, 11, 10

Minimum 3 Lower quartile 7 Median 11 Upper quartile 13 Maximum 15

Does it matter that the groups aren't even?

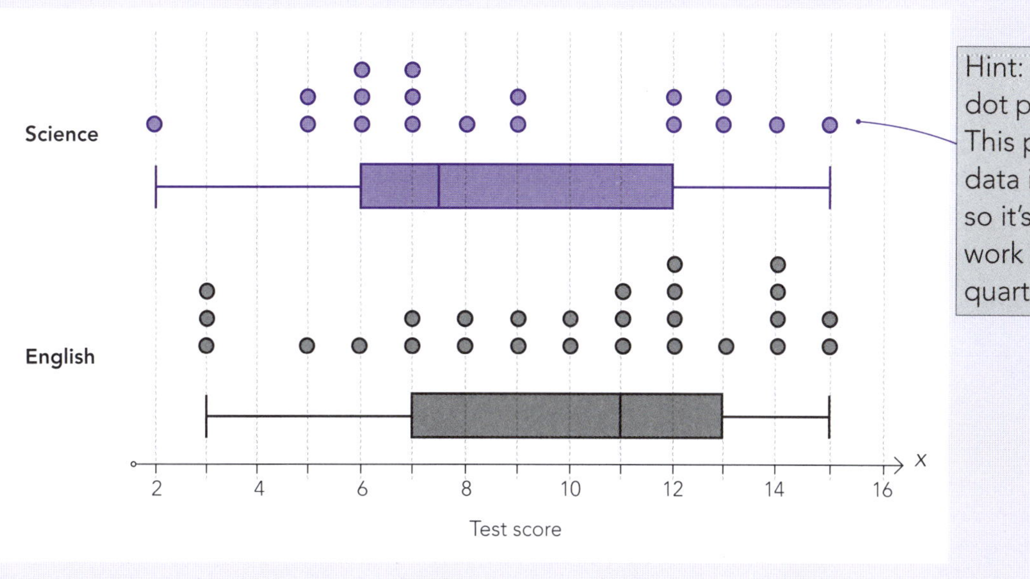

Hint: Draw the dot plot first. This puts your data in order so it's easy to work out the quartiles, etc.

ISBN: 9780170370417

Draw dot plots and box plots for the following data sets.

1 **1, 1, 2, 5, 6, 7, 9, 10, 10, 11, 12, 12, 13, 13, 13, 13, 14, 14, 15, 15, 15, 16, 17, 18, 20, 20, 20, 20, 21, 21**

Minimum ____ Lower quartile ____ Median ____ Upper quartile ____ Maximum ____

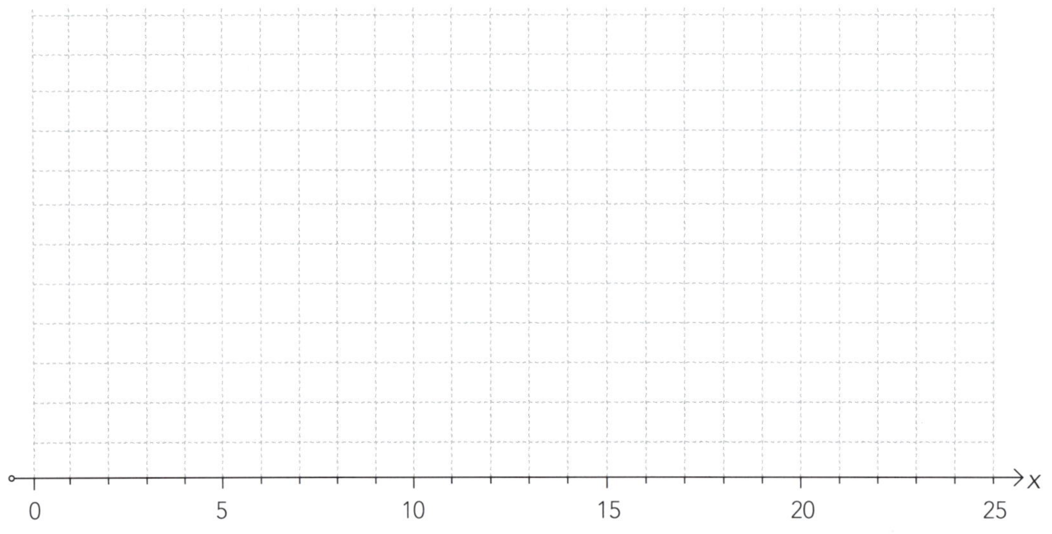

2 **4, 12, 5, 7, 1, 6, 8, 5, 4, 2, 14, 12, 14, 9, 6, 21, 24, 1, 3, 20, 15**

Minimum ____ Lower quartile ____ Median ____ Upper quartile ____ Maximum ____

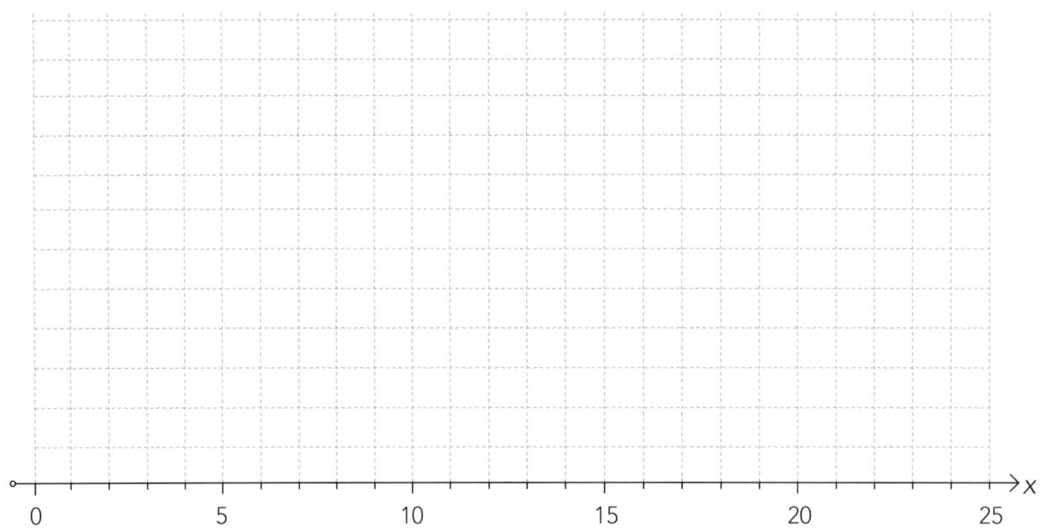

  ISBN: 9780170370417

3 Hours spent playing sport each week.

Year 10: 0, 0, 1, 1, 1, 2, 2, 2, 2, 3, 3, 3, 3, 3, 4, 7, 8, 12

Year 12: 0, 0, 0, 0, 0, 1 ,1, 1, 2, 2, 2, 4, 4, 5, 5, 8, 10

Note: You will need **2** dot plots and **2** box plots for this question.

Year 10:
Minimum ____ Lower quartile ____ Median ____ Upper quartile ____ Maximum ____

Year 12:
Minimum ____ Lower quartile ____ Median ____ Upper quartile ____ Maximum ____

ISBN: 9780170370417

4 Number of apps on their phones.

Girls: 2, 5, 8 ,4, 1, 9, 12, 21, 15, 17, 6, 8, 9, 5, 3, 20, 24, 17, 16, 17

Boys: 5, 2, 9, 15, 15, 15, 17, 21, 3, 15, 12, 15, 15, 6, 17, 4, 22, 7, 12

____

____

Girls:
Minimum ____ Lower quartile ____ Median ____ Upper quartile ____ Maximum ____

Boys:
Minimum ____ Lower quartile ____ Median ____ Upper quartile ____ Maximum ____

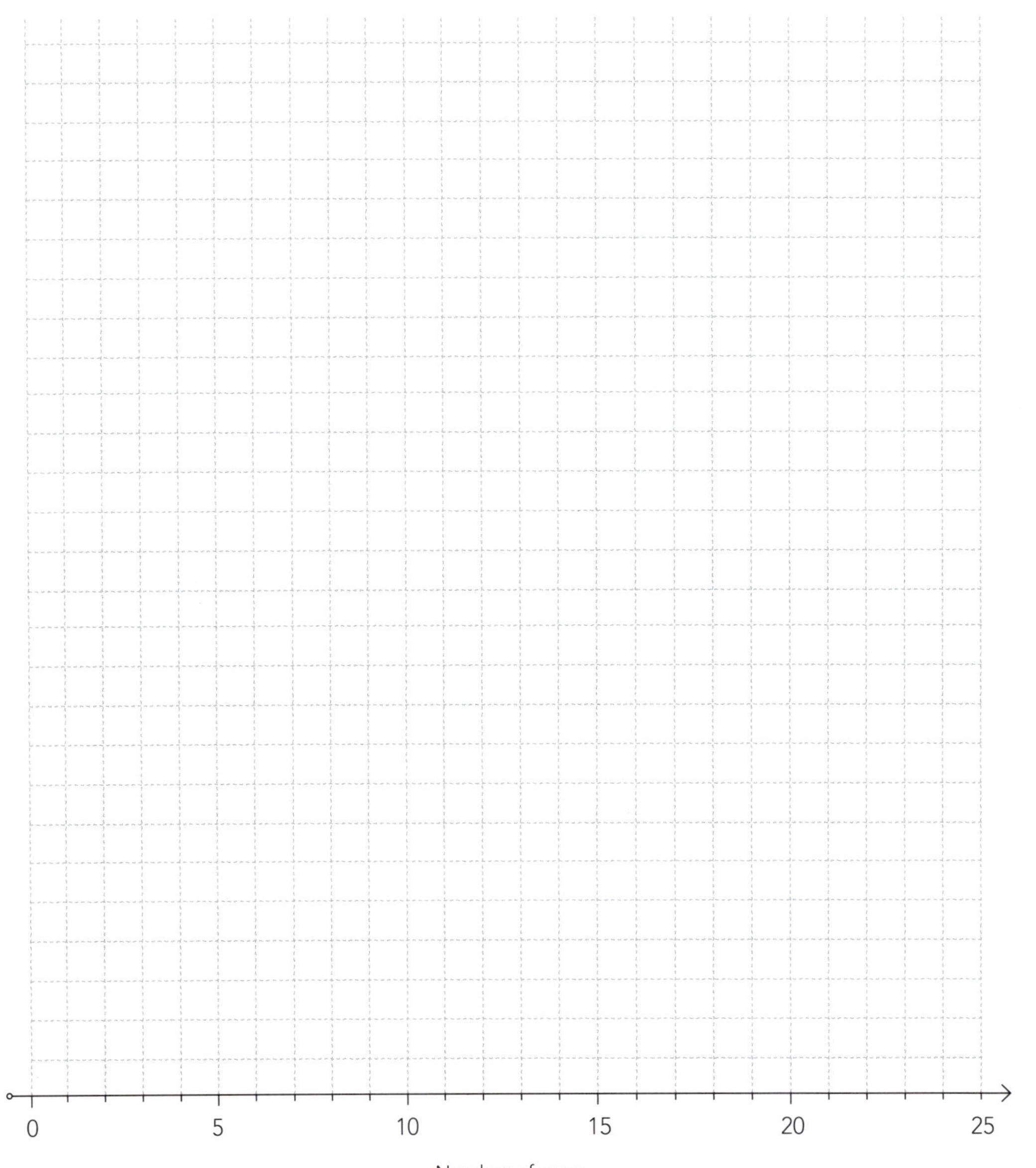

 ISBN: 9780170370417

Combining dot plots, box plots and means

The first three examples from pages 12–15 have the same data as below. The answers have been put on the same graph so you can see the relationship between them.

1

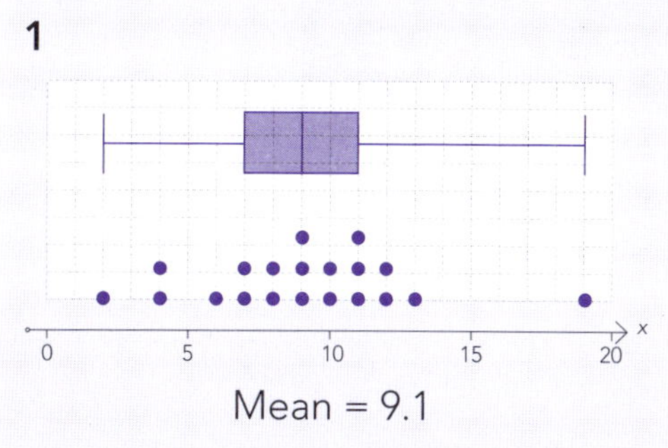

Mean = 9.1

- The mean (9.1) and the median (9) are very close because the data is distributed fairly symmetrically around the mean.
- Notice that the length of the whisker on the right is long because of just one value.
- The gaps within the box are much smaller than the lengths of the whiskers because the data is clustered in the centre.

2

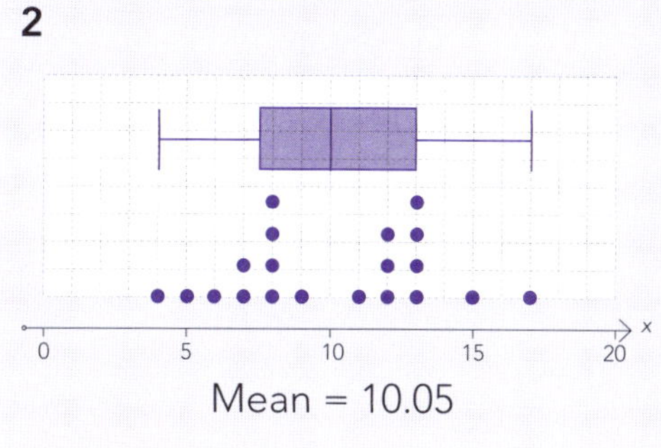

Mean = 10.05

- The data is fairly symmetrically spread around the median, so the median (10) and the mean (10.05) are close.
- The dot plot shows that this data is 'bimodal' — it has two modes. This is **not** shown by the box plot.
- Because of the gap in the middle of the data, the minimum, lower quartile, median, upper quartile and maximum are fairly evenly spread.

3

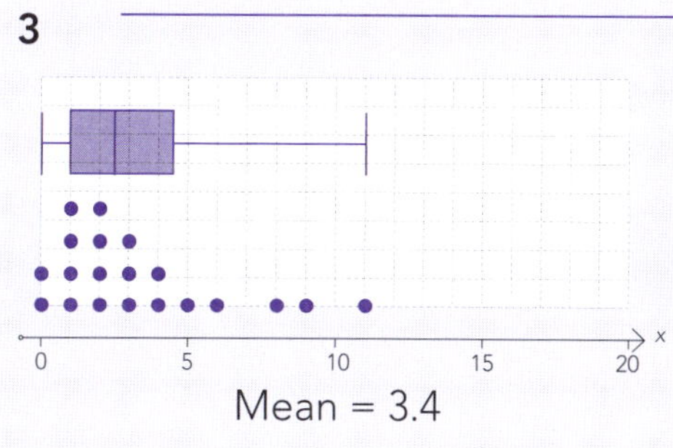

Mean = 3.4

- Most of the data is clumped on the left, which means the gaps between the minimum, lower quartile and median are small.
- The mean (3.4) is bigger than the median (2.5) because the 'tail' of bigger numbers on the right increases the mean, but not the median.
- The data is **skewed to the right**.

Match the dot plots with the box plots and delete words to make a true statement about the mean.

1

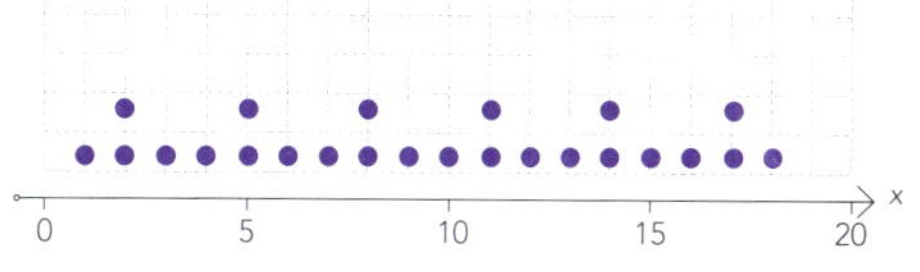

Mean is above/below/about the same as the median.

a

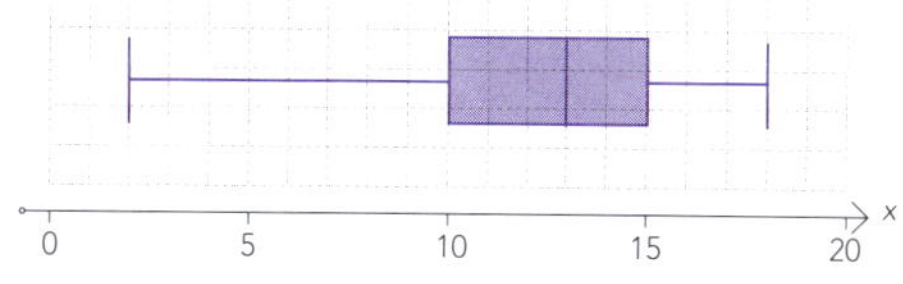

2

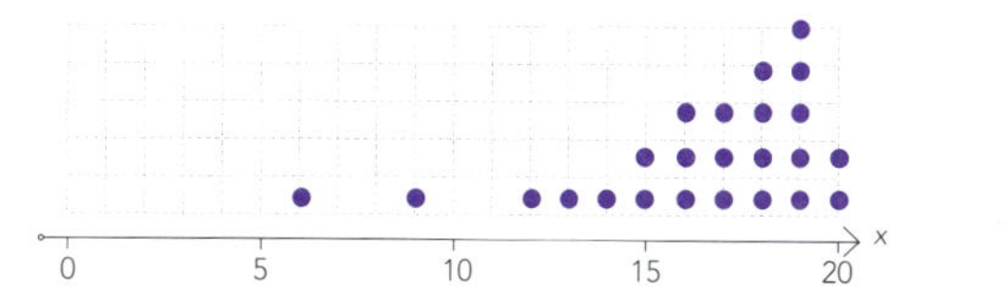

Mean is above/below/about the same as the median.

b

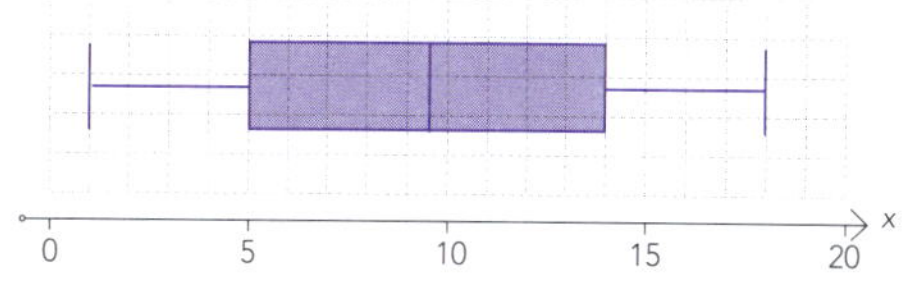

3

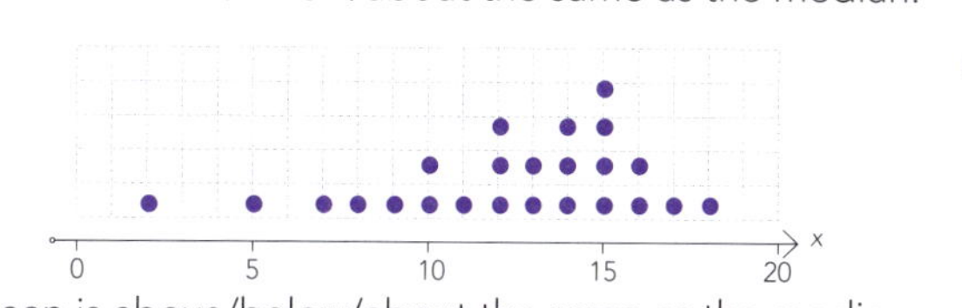

Mean is above/below/about the same as the median.

c

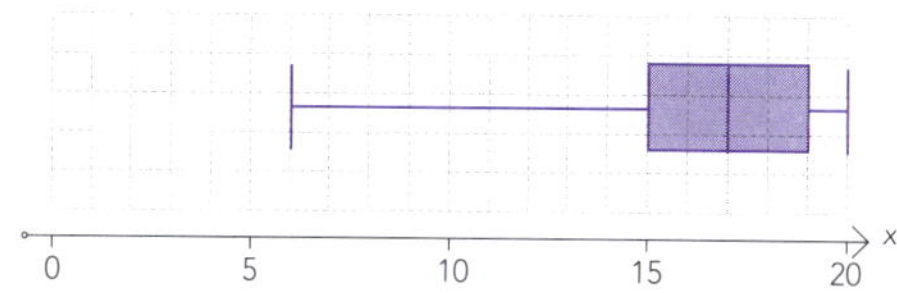

ISBN: 9780170370417

Box plot interpretation

1 Explain why a box plot might look like this.

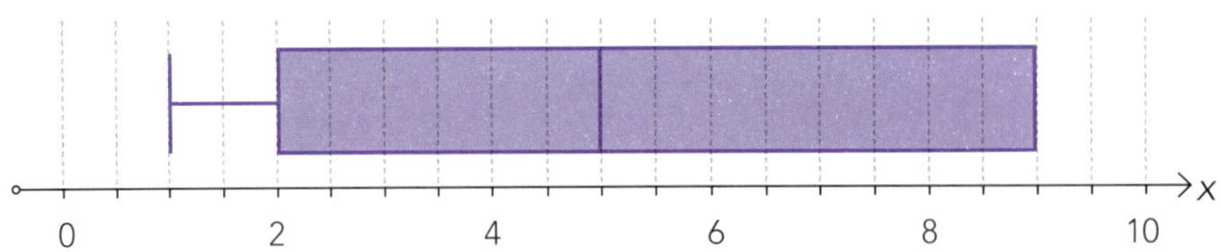

2 Explain why a box plot might look like this.

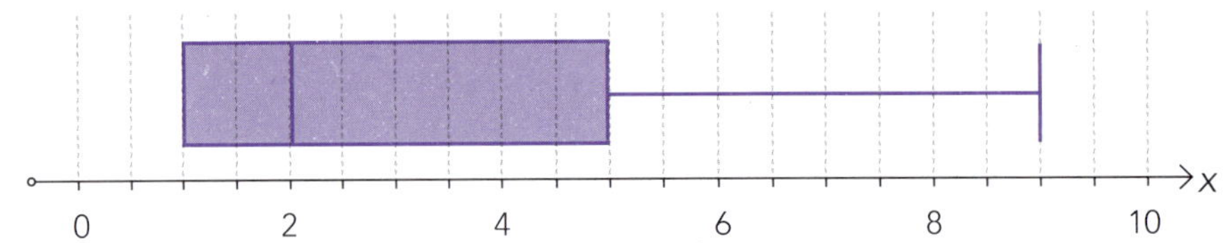

3 Explain why a box plot might look like this.

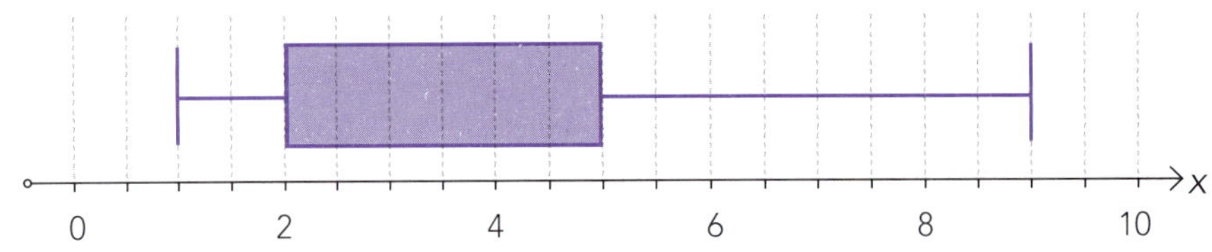

Thinking about dot and box plots

1 What does a dot plot show us that a box and whisker doesn't?

2 What does a box and whisker show us that a dot plot doesn't?

3 What do both show?

ISBN: 9780170370417

Planning an investigation

Writing questions

The question needs:

- to compare two groups, e.g. male/female
- to indicate direction, e.g. one is bigger/longer/smaller, etc. than the other
- to use numeric data
- if possible, to mention the population.

This is a good structure:

I would like to know if the ________ [Variable] of ________ [Group A] tend to be larger/smaller than the ________ [Variable] of ________ [Group B] from ________ [Population].

Examples:

1 I would like to know if the **weights** of **dogs that sleep outside** tend to be greater than the **weights** of **dogs that sleep inside** for **dogs that live in Christchurch**.

2 I would like to know if the **heights** of **Year 9 boys** tend to be greater than the **heights** of **Year 9 girls** at **Paradise High School**.

If you feel confident, you can reorganise the sentences.

Examples:

1 I would like to know if, in Christchurch, dogs that sleep outside tend to be heavier than dogs that sleep inside.

2 I would like to know if Year 9 boys tend to be taller than Year 9 girls at Paradise High School.

Good and bad questions:

What is the difference between boys' and girls' bag weights?

It doesn't have a direction indication (larger or smaller).
No population is mentioned.

I would like to know if the bag weight for Year 12 boys tends to be greater than the bag weight for Year 12 girls at Paradise High School.

ISBN: 9780170370417

Are these questions acceptable or not?

	Yes/No	Why?	Make it better
Are 13-year-old boys taller than 13-year-old girls?			
What is the relationship between a student's arm span and height?			
I would like to know how the heights of Year 11 boys compare with the heights of Year 11 girls at Paradise High School.			
I would like to know if the bag weight of Year 11 girls tends to be greater than the bag weight of Year 12 girls.			
What is the difference in hair of a Year 9 class at Paradise High School?			
Do boys take longer to get to school than girls?			
I would like to know the difference between how boys and girls at Paradise High School get to school.			

ISBN: 9780170370417

This is a sample of the cats from the town of Flockton.

Sex	Breed	Age (years)	Weight (kg)	Sleeps	Litter size	Collar
Female	Devon Rex	3.4	4.9	Inside	4	Yes
Male	Moggy	2.6	3.1	Outside	3	No
Male	Coon	0.7	2.1	Inside	3	No
Female	Siamese	1.2	2.4	Inside	2	Yes
Female	Burmese	10.2	4.9	Inside	5	No
Female	Moggy	5.6	3.7	Inside	6	No
Female	Persian	4.3	3.8	Inside	3	Yes
Male	Moggy	12.4	6.5	Outside	4	Yes
Female	Bengal	6.0	4.7	Outside	1	No
Male	Siamese	0.5	1.8	Inside	2	Yes
Male	Persian	4.8	4.9	Inside	7	No
Male	Moggy	7.2	5.3	Outside	4	No
Female	Burmese	9.1	5.9	Inside	7	Yes
Male	Birman	5.2	5.6	Inside	2	Yes
Female	Moggy	8.3	4.2	Inside	5	Yes
Male	Manx	2.6	3.7	Outside	3	No
Male	Moggy	4.8	5.1	Inside	7	No
Female	Moggy	1.0	2.2	Outside	2	No
Female	Burmese	6.9	3.8	Outside	4	Yes
Female	Siamese	11.3	4.9	Inside	5	Yes
Male	Moggy	9.7	5.8	Inside	2	No
Female	Burmese	2.3	4.0	Inside	5	Yes

Write three appropriate questions for this data set.

1 ______________________________

____

2 ______________________________

____

3 ______________________________

____

ISBN: 9780170370417

Collecting data

A **census** is where we collect data from **every** member of the population. This is usually too expensive, or impossible.

So, when collecting data we nearly always have to take a **sample** from our population.

We then use sample data to make **inferences** about the population.

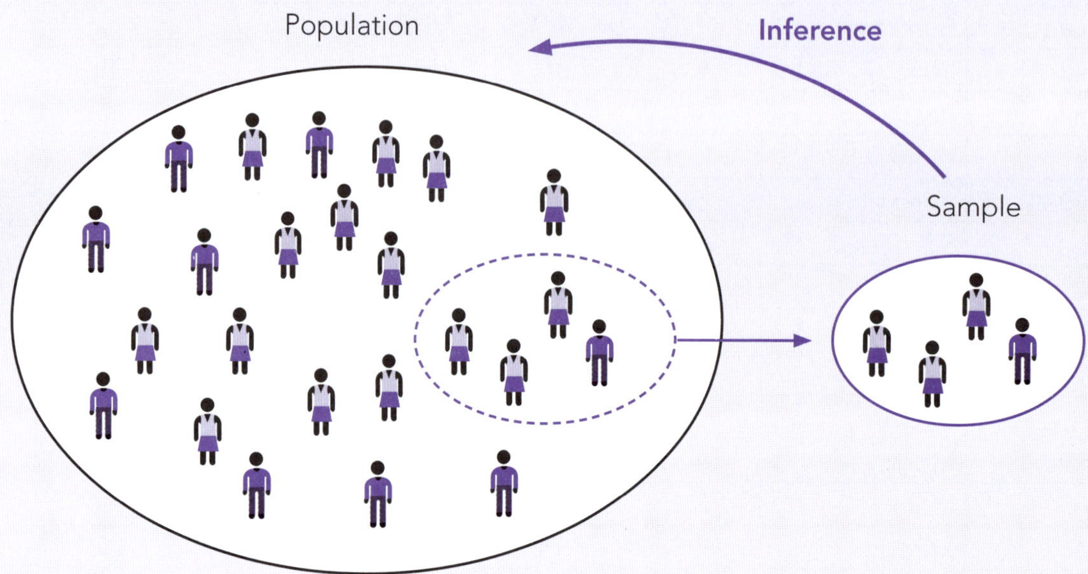

There are two important questions to ask when considering data from a sample:

1 **Is the sample big enough?**
A sample of 30 is generally considered enough for most purposes. However, the bigger the sample, the more confidence you can have of your findings.

2 **Is the data biased?**
Bias occurs when some members of the population are more likely than others to be selected for the sample, so the sample does not truly represent the population.

Examples:

- Phoning a radio station, filling in a form in a magazine or going to a website to give feedback: these are called 'self-selected' samples because the subject decides whether they will be sampled. Only those with an interest in the topic of the survey will be in the sample.
- The survey in the mall: this collects data only from those who go to malls, those who have time to answer, and there is an element of self-selection.
- Telephone surveys: are often done on land lines, so mostly older people and people who don't often move house are surveyed. These also tend to be self-selected because many people choose not to respond.
- Surveying just your friends.

ISBN: 9780170370417

State whether the following samples might be biased, and why.

1 In order to find the views of Year 10 students, the dean surveys one class of Year 10 students.

2 In order to find the views of Paradise High School students, students walking past the senior common room are surveyed.

3 In order to find the views of Paradise High School students, every tenth student passing through the gate before 8.30 a.m. is surveyed.

4 In order to find the views of Paradise High School students, every tenth student on the school roll is surveyed.

5 In order to find the views of Paradise High School students about what food should be sold at the canteen, students queuing at the canteen at lunchtime are surveyed.

6 In order to find the views of Paradise High School students about what food should be sold at the canteen, response forms are left outside the student office, along with a drop box.

7 In order to find the views of Paradise High School students, all students with a family name starting with a particular letter of the alphabet are selected.

8 In order to find the views of Paradise High School students, 10 students are selected randomly from each year level.

ISBN: 9780170370417

Sampling

In order to avoid bias, there needs to be a system for sampling the population so that **every member of the population has an equal chance of being selected**.

Some systems of sampling

1 **Systematic sampling:** this involves taking every nth member of the population.
Examples:
- Selecting every fifth sheep as they pass through a narrow gap.
- Giving every member of the population a number, and then selecting every tenth individual, or those with odd numbers.
- Selecting every fifth car in a car park.
- Selecting every twentieth person on the school roll.

2 **Random sampling**
Examples:
- Writing names on equal-sized pieces of paper and drawing them from a hat.
- Writing every number on a ball and using a machine to select the balls, as in Lotto.
- Giving every member of the population a number and using random numbers to decide who will be in the sample.
- Using random numbers to decide who will be sampled from the electoral roll.

How to use random numbers

1 Give all the members of your population a number.
2 Find the 'Random Number' button on your calculator.
3 Usually, once in Random mode, if you press the '=' sign you will get a three-digit random number.
4 Decide how big your random number will need to be.
- If you have 99 or fewer members, you will need two-digit random numbers.
- If you have between 100 and 999 members, you will need three-digit random numbers. And so on.

5 Assume there are 60 in your population, so you need two-digit random numbers. This means that you use **only** the first two digits after the decimal point.

Random number	Number selected	Explanation
0.**20**3	20	
0.**58**2	58	
0.**71**0	none	Only 60 in the population, so ignore this random number.
0.**04**5	4	Do not ignore the 0.
0.**00**3	none	Number 0 does not exist.
0.**10**0	10	

6 Keep going until you have the number required for the sample.

 ISBN: 9780170370417

This is a sample of the cats from the town of Flockton.

	Sex	Breed	Age (years)	Weight (kg)	Sleeps	Litter size	Collar
1	Female	Devon Rex	3.4	4.9	Inside	4	Yes
2	Male	Moggy	2.6	3.1	Outside	3	No
3	Male	Coon	0.7	2.1	Inside	3	No
4	Female	Siamese	1.2	2.4	Inside	2	Yes
5	Female	Burmese	10.2	4.9	Inside	5	No
6	Female	Moggy	5.6	3.7	Inside	6	No
7	Female	Persian	4.3	3.8	Inside	3	Yes
8	Male	Moggy	12.4	6.5	Outside	4	Yes
9	Female	Bengal	6.0	4.7	Outside	1	No
10	Male	Siamese	0.5	1.8	Inside	2	Yes
11	Male	Persian	4.8	4.9	Inside	7	No
12	Male	Moggy	7.2	5.3	Outside	4	No
13	Female	Burmese	9.1	5.9	Inside	7	Yes
14	Male	Birman	5.2	5.6	Inside	2	Yes
15	Female	Moggy	8.3	4.2	Inside	5	Yes
16	Male	Manx	2.6	3.7	Outside	3	No
17	Male	Moggy	4.8	5.1	Inside	7	No
18	Female	Moggy	1.0	2.2	Outside	2	No
19	Female	Burmese	6.9	3.8	Outside	4	Yes
20	Female	Siamese	11.3	4.9	Inside	5	Yes
21	Male	Moggy	9.7	5.8	Inside	2	No
22	Female	Burmese	2.3	4.0	Inside	5	Yes
23	Male	Moggy	4.3	6.3	Inside	3	Yes
24	Female	Bengal	2.1	3.5	Outside	4	No
25	Female	Persian	4.0	4.1	Outside	2	No
26	Male	Devon Rex	1.9	3.3	Outside	1	No
27	Female	Moggy	2.6	3.8	Inside	7	Yes
28	Male	Birman	5.2	4.9	Inside	5	No
29	Male	Siamese	0.3	1.0	Inside	4	No
30	Male	Moggy	2.8	2.6	Inside	2	Yes

(Continued on next page.)

ISBN: 9780170370417

	Sex	Breed	Age (years)	Weight (kg)	Sleeps	Litter size	Collar
31	Male	Moggy	7.2	4.7	Inside	4	No
32	Female	Burmese	3.6	4.1	Outside	4	No
33	Female	Burmese	6.1	3.7	Inside	6	No
34	Male	Manx	2.3	3.2	Outside	4	No
35	Male	Moggy	4.7	4.3	Inside	2	Yes
36	Female	Moggy	1.7	2.6	Inside	1	Yes
37	Male	Coon	2.4	4.6	Outside	3	No
38	Female	Moggy	5.3	4.3	Outside	4	No
39	Female	Siamese	5.1	3.5	Inside	2	No
40	Male	Moggy	9.3	5.1	Inside	3	Yes
41	Female	Siamese	4.6	3.6	Inside	3	No
42	Female	Burmese	6.1	4.2	Inside	5	No
43	Male	Moggy	7.8	5.1	Outside	2	Yes
44	Male	Persian	2.0	4.6	Inside	1	Yes
45	Male	Birman	3.1	4.2	Inside	4	Yes
46	Female	Moggy	1.5	3.2	Outside	3	No
47	Female	Burmese	5.4	3.7	Inside	2	No
48	Female	Burmese	2.2	3.0	Inside	5	No
49	Male	Moggy	5.7	4.3	Inside	4	Yes
50	Male	Burmese	3.1	4.1	Outside	6	Yes
51	Female	Birman	2.8	3.7	Inside	4	No
52	Female	Siamese	6.7	3.5	Inside	3	No
53	Male	Bengal	3.7	6.2	Inside	5	Yes
54	Female	Moggy	1.6	2.9	Outside	2	No
55	Male	Moggy	0.9	1.5	Inside	6	Yes
56	Male	Moggy	3.4	4.1	Inside	4	No
57	Female	Burmese	2.4	3.2	Outside	4	No
58	Male	Siamese	7.1	4.6	Outside	3	No
59	Female	Moggy	6.3	4.1	Inside	6	No
60	Female	Moggy	2.7	3.6	Inside	2	Yes

ISBN: 9780170370417

Sampling variation

1 Select and highlight a sample of **30** cats from the table on pages 27–28, and pick an appropriate variable (e.g. weight) to calculate the required statistics.

Space for working:

Minimum ____ Lower quartile ____ Median ____ Upper quartile ____ Maximum ____

2 Draw box and dot plots for your results.

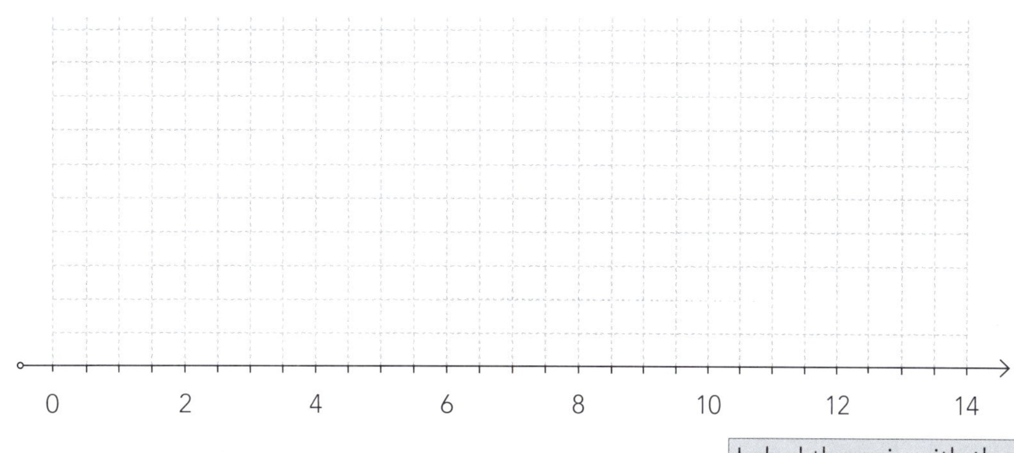

____

Label the axis with the variable (and units) you chose.

3 Get together with others who chose the same variable as you did.

Are your graphs the same or different? ______________________

Explain why this is.

____

____

ISBN: 9780170370417

4 Mason selected all the cats that sleep inside. What's wrong with this?

____

____

5 Select and highlight different sample of **10** cats from the table on pages 27–28, and use the same variable (e.g. weight) to calculate the required statistics.

Space for working:

Minimum ____ Lower quartile ____ Median ____ Upper quartile ____ Maximum ____

6 Draw box and dot plots for your results.

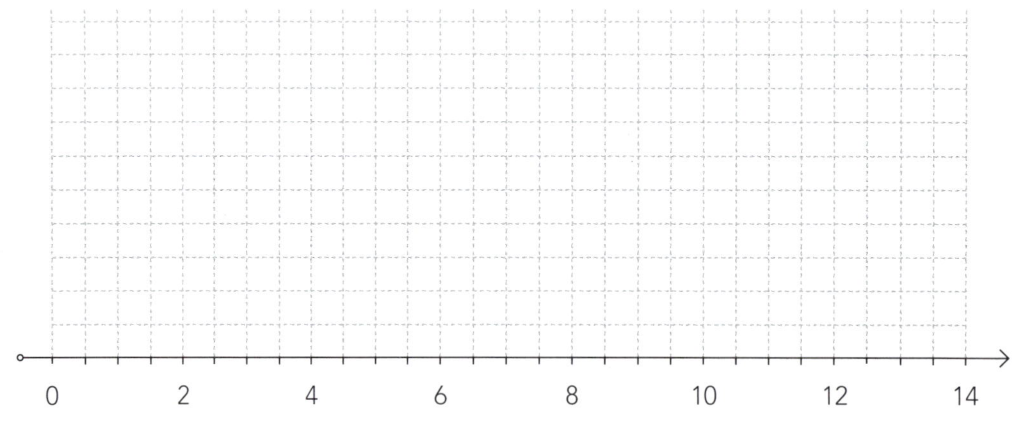

____

7 Get together again with others who chose the same variable as you did. How do the graphs from your samples of 30 compare with those from the samples of 10?

____

Explain why this is.

____

____

 ISBN: 9780170370417

Sampling variation summary

1 Use the words in the column below right to complete this page.

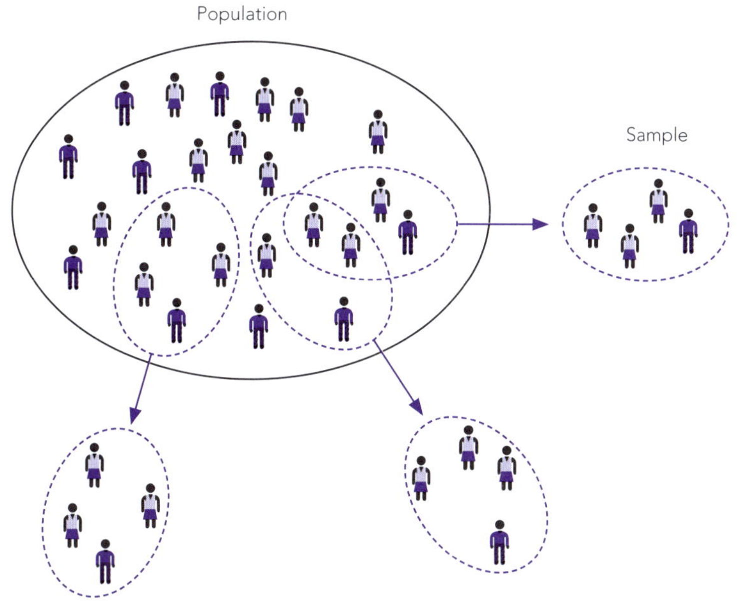

Every sample taken from a ______________________ will be different.

If I repeated my sampling and analysis, I would get

______________________ results.

If I took a **bigger** sample:

- My results would be ______________________.
- The mean and ______________________ would probably be ______________________ to the population ______________________ and median.
- The IQR would probably be ______________________ to the population IQR.

If I took a **smaller** sample:

- My results would be ______________________.
- The mean and ______________________ would probably be ______________________ from the population ____________ and median.
- The IQR would probably be ______________________ from the population IQR.

further
mean
closer
different
mean
population
more reliable
closer
less reliable
median
further
median

Unless you do a census, you will *never* know the population mean, median or IQR.

ISBN: 9780170370417

Describing and comparing features

This involves writing comparative sentences to describe the differences and similarities between two groups. You need to:

- compare the centres
- describe and compare the shape
- compare the positions and sizes of the spreads
- discuss where the data overlaps
- describe and discuss any unusual features.

Centres

1 Comparing centres using statistics

You will be required to calculate statistics of two groups and compare them.

Example: A group of Year 11 students got their results back from two recent tests. Both tests were out of 20 marks.

Science: **2, 5, 7, 9, 8, 12, 14, 6, 13, 15, 7, 9, 6, 5, 7, 13, 6, 12**

Remember you need to put them in order.

English: **3, 12, 14, 5, 14, 13, 12, 14, 15, 3, 3, 7, 8, 9, 6, 10, 8, 7, 12, 11, 9, 11, 15, 14, 12, 11, 10**

Science: Median 7.5 Mean $8.\dot{6}$

English: Median 11 Mean 9.926

Is this group of Year 11 students better at Science or English? Why?

What do you see?
The median English mark is 3.5 more than the median Science mark. The mean for English (9.926) is also higher than the mean for Science ($8.\dot{6}$).

What does this mean for the sample?
This means that marks for this group (or sample) of students **tend** to be higher in English than in Science.

We cannot say they '**are**' higher, because not all English marks are higher than Science marks.

Does it matter that the two groups have different numbers of marks?

ISBN: 9780170370417

Calculate the statistics for the following data sets and make a statement:

1 A group of students recorded how many minutes it takes them to get ready for school in the morning (min). Here are the results:

Boys: 3, 4, 12, 16, 17, 17, 19, 20, 21, 22, 22, 25, 45
Girls: 13, 14, 15, 15, 18, 18, 25, 26, 26, 27, 28, 32, 33, 34, 52

Boys: Median ______ Mean ______

Girls: Median ______ Mean ______

Which gender takes longer to get ready in the morning? Justify your answer.

____

____

2 A group of Year 9 students measured their heights (cm). Here are the results:

Boys: 158, 164, 163, 157, 155, 168, 159, 163, 157, 155, 156, 170
Girls: 155, 153, 162, 154, 152, 148, 157, 160, 145, 155

____

____

Boys: Median ______ Mean ______

Girls: Median ______ Mean ______

Are these Year 9 boys or girls taller? Why?

____

____

3 A group of students recorded the time it took them to get to school (min). Here are the results:

Boys: 16, 19, 9, 14, 31, 12, 24, 16, 19, 24, 31, 26, 34, 2, 13, 17, 5, 6
Girls: 17, 21, 6, 15, 32, 11, 3, 12, 17, 16, 33, 28, 24, 22, 19, 1, 6, 8, 14

____

____

Boys: Median ______ Mean ______

Girls: Median ______ Mean ______

Did the boys or girls take longer to get to school? Why?

____

____

ISBN: 9780170370417

2 Comparing centres using dot plots

We can also plot two data sets in order to compare distributions.

Example: A group of Year 11 students got their results back from two recent tests.

Science: **2, 5, 7, 9, 8, 12, 14, 6, 13, 15, 7, 9, 6, 5, 7, 13, 6, 12**

English: **3, 12, 14, 5, 14, 13, 12, 14, 15, 3, 3, 7, 8, 9, 6, 10, 8, 7, 12, 11, 9, 11, 15, 14, 12, 11, 10**

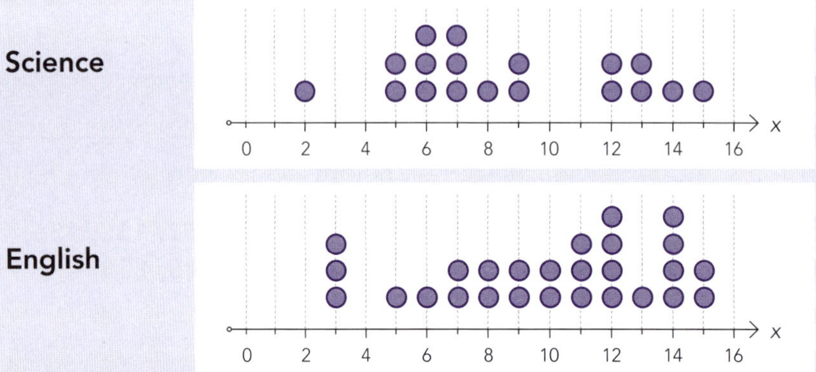

Is this group of Year 11 students better at Science or English? Why?
The dot plot suggests that the Year 11 students in these samples tend to do better in English than Science because the dots for English are further up to the right.

Create dot plots for these data sets and answer the questions.

1 Time spent doing homework on Wednesday night.

Boys: 0, 0, 2, 2, 3, 4, 5, 5, 5, 9, 9, 10, 11, 12, 13, 15, 20, 22, 24, 31
Girls: 0, 4, 4, 6, 8, 9, 10, 10, 12, 12, 12, 14, 16 ,16 ,17, 19, 21, 22, 23, 25, 25, 30

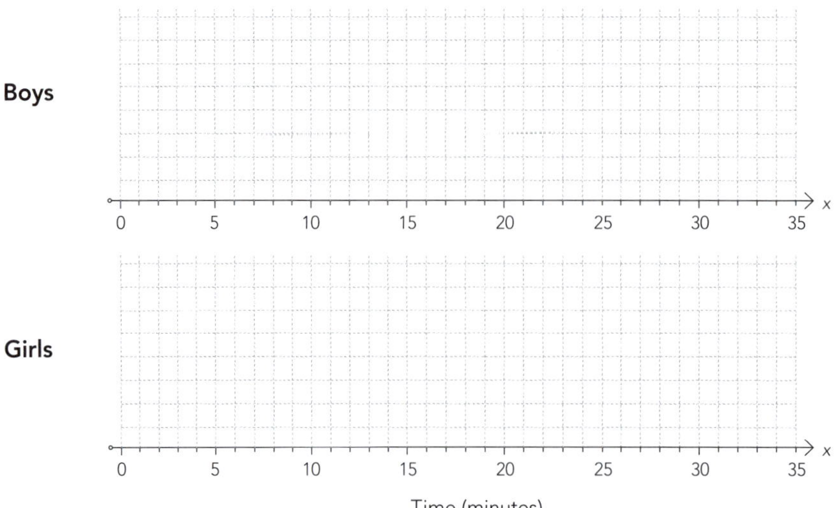

Who spent longer doing homework on a Wednesday night? Justify your answer.

____

____

ISBN: 9780170370417

2 Number of days absent in a term for 20 randomly selected Year 9 and 11 students.

Year 9: 20, 15, 10, 0, 1, 5, 0, 14, 12, 0, 13, 9, 5, 0, 12, 11, 6, 8, 0, 0

Year 11: 0, 0, 1, 3, 1, 0, 2, 4, 10, 22, 0, 0, 14, 5, 1, 2, 7, 0, 0, 12

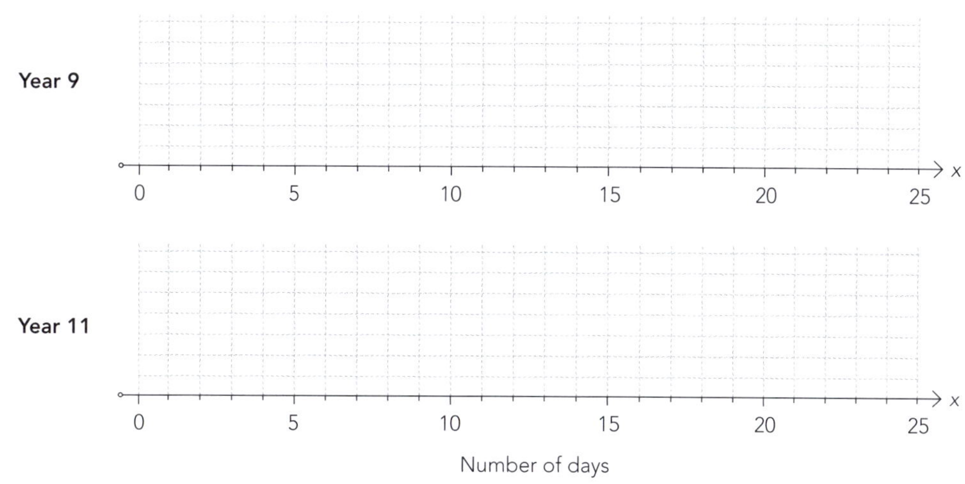

Which year group has more absences? Justify your answer.

3 Number of pairs of shoes owned by 20 Year 13 boys and girls.

Boys: 3, 1, 2, 4, 5, 2, 2, 10, 5, 6, 8, 10, 3, 1, 4, 2, 2, 3, 3, 4

Girls: 4, 6 ,8 ,14, 7, 5, 6, 24, 12, 15, 18, 14, 12, 9, 8, 7, 2, 6, 6, 10

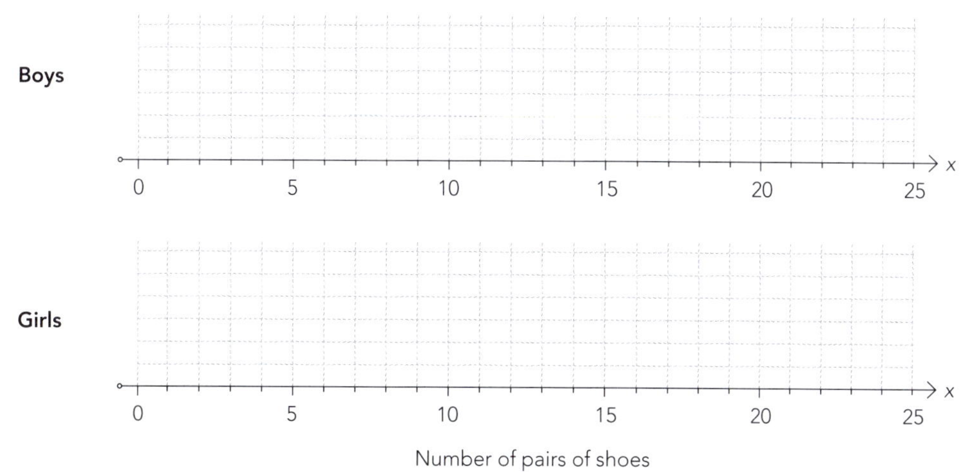

Who owns more shoes, boys or girls? Justify your answer.

ISBN: 9780170370417

3 Comparing centres using box plots

A group of Year 11 students got their results back from two recent tests.

Science: **2, 5, 7, 9, 8, 12, 14, 6, 13, 15, 7, 9, 6, 5, 7, 13, 6, 12**

English: **3, 12, 14, 5, 14, 13, 12, 14, 15, 3, 3, 7, 8, 9, 6, 10, 8, 7, 12, 11, 9, 11, 15, 14, 12, 11, 10**

Science Lower quartile 6 Median 7.5 Upper quartile 12

English Lower quartile 7 Median 11 Upper quartile 13

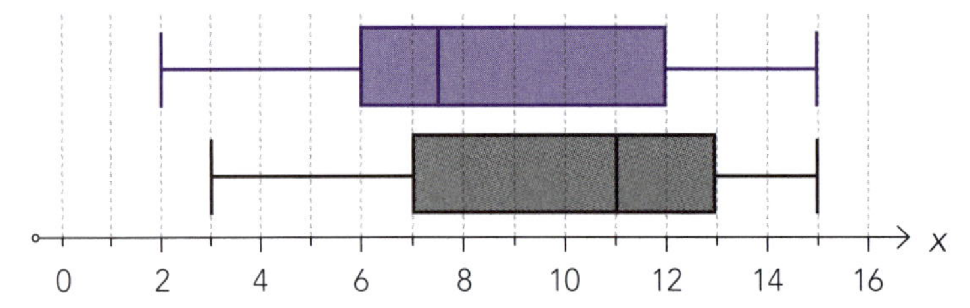

Is this group of Year 11 students better at Science or English? Why?
This group of Year 11 students tended to do better at English, because both the box and the median are further to the right.

Draw the box plots for these data sets and answer the questions.

1 Time spent doing homework on Monday night.

Boys: 0, 0, 2, 2, 3, 4, 5, 5, 5, 9, 9, 10, 11, 12, 13, 15, 20, 22, 24, 31
Girls: 0, 4, 4, 6, 8, 9, 10, 10, 12, 12, 12, 14, 16 ,16 ,17, 19, 21, 22, 23, 25, 25, 30

Boys:
Minimum ____ Lower quartile ____ Median ____ Upper quartile ____ Maximum ____

Girls:
Minimum ____ Lower quartile ____ Median ____ Upper quartile ____ Maximum ____

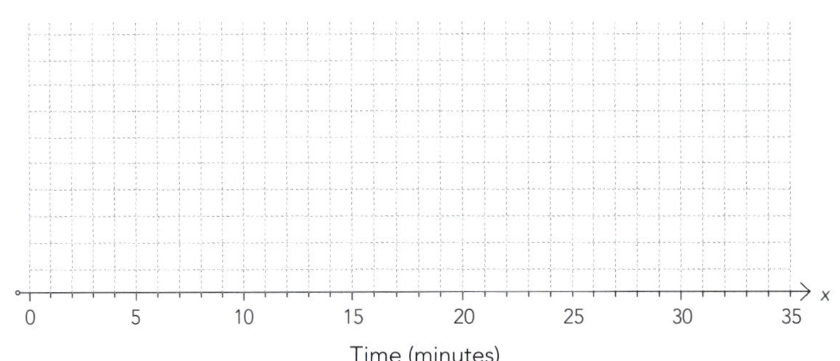

Who does more homework on a Monday night? Justify your answer.

___

___

ISBN: 9780170370417

2 Number of days absent in a term for 20 randomly selected Year 9 and 11 students.

Year 9: 20, 15, 10, 0, 1, 5, 0, 14, 12, 0, 13, 9, 5, 0, 12, 11, 6, 8, 0, 0

____

Year 11: 0, 0, 1, 3, 1, 0, 2, 4, 10, 22, 0, 0, 14, 15, 1, 2, 7, 0, 0, 12

____

Year 9:

Minimum ____ Lower quartile ____ Median ____ Upper quartile ____ Maximum ____

Year 11:

Minimum ____ Lower quartile ____ Median ____ Upper quartile ____ Maximum ____

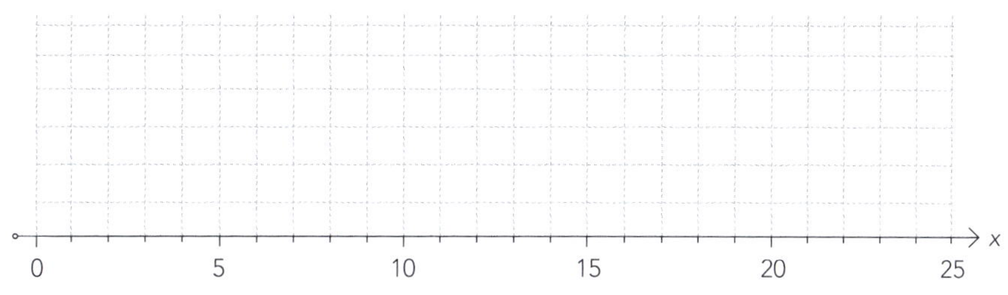

Which year group has more absences? Justify your answer.

____

____

3 Number of pairs of shoes owned by 20 Year 13 boys and girls.

Boys: 3, 1, 2, 4, 5, 2, 2, 10, 5, 6, 8, 10, 3, 1, 4, 2, 2, 3, 3, 4

____

Girls: 4, 6 ,8 ,14, 7, 5, 6, 24, 12, 15, 18, 14, 12, 9, 8, 7, 2, 6, 6, 10

____

Boys:

Minimum ____ Lower quartile ____ Median ____ Upper quartile ____ Maximum ____

Girls:

Minimum ____ Lower quartile ____ Median ____ Upper quartile ____ Maximum ____

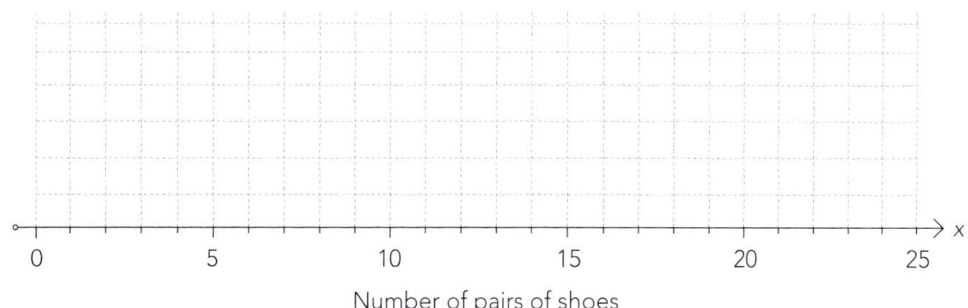

Who owns more shoes, boys or girls? Justify your answer.

____

____

ISBN: 9780170370417

4 Minutes spent on Science homework last night.

Year 9: 0, 1, 8, 12, 9, 10, 14, 8, 11, 8, 8, 0, 12, 9, 8, 8

Year 11: 2, 2, 5, 0, 0, 4, 15, 24, 0, 0, 18, 12, 6, 8, 0, 12

Year 9:
Minimum ____ Lower quartile ____ Median ____ Upper quartile ____ Maximum ____

Year 11:
Minimum ____ Lower quartile ____ Median ____ Upper quartile ____ Maximum ____

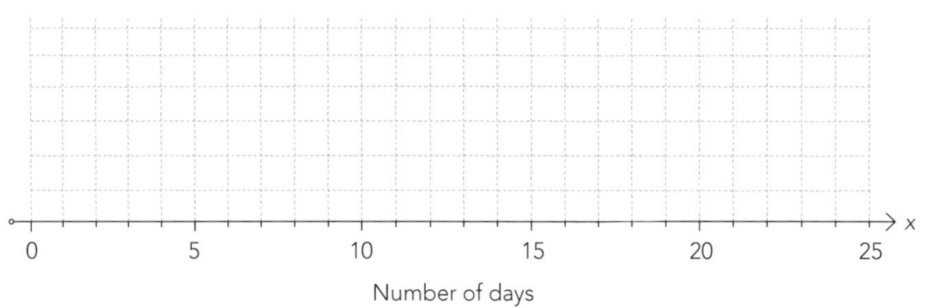

Which year spent longer on Science homework last night? Justify your answer.

5 Marks out of 24 in Geometry and Algebra tests.

Geometry: 24, 15, 24, 24, 18, 22, 14, 12, 23, 18, 23, 11, 10, 20, 17, 24, 24

Algebra: 15, 23, 22, 10, 22, 20, 24, 12, 22, 19, 23, 16, 22, 18, 19, 22, 24

Geometry:
Minimum ____ Lower quartile ____ Median ____ Upper quartile ____ Maximum ____

Algebra:
Minimum ____ Lower quartile ____ Median ____ Upper quartile ____ Maximum ____

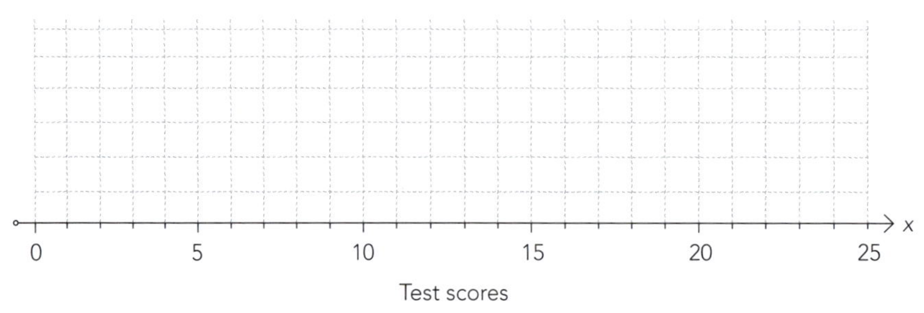

In which subject did the class get higher marks? Justify your answer.

ISBN: 9780170370417

Shape

1 Describing shapes

The samples need to be large enough to show the shape.

Normal distribution
(hill/mound shape, symmetrical, bell-shaped curve)

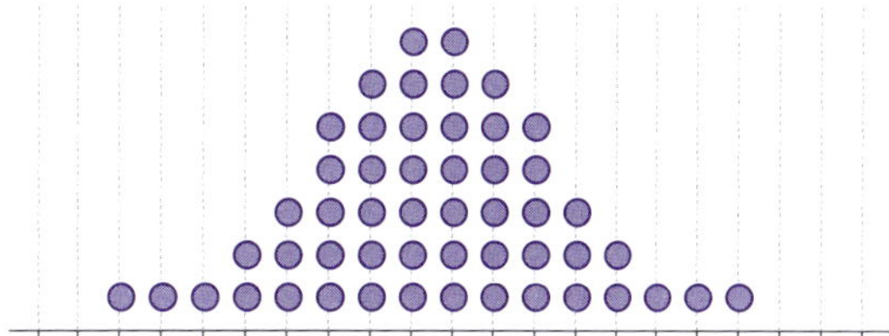

Right skew
(the tail is on the right-hand side)

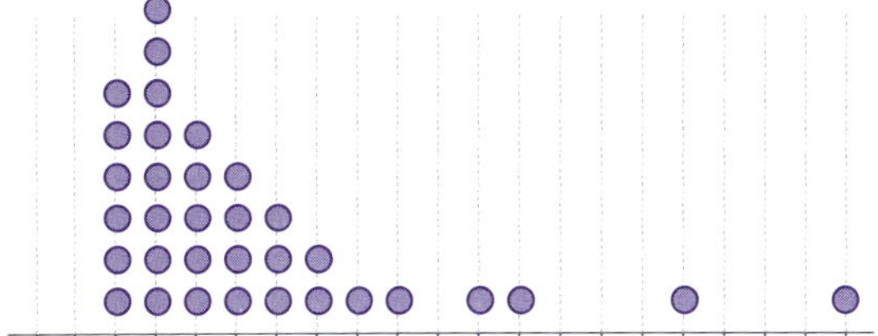

Left skew
(the tail is on the left-hand side)

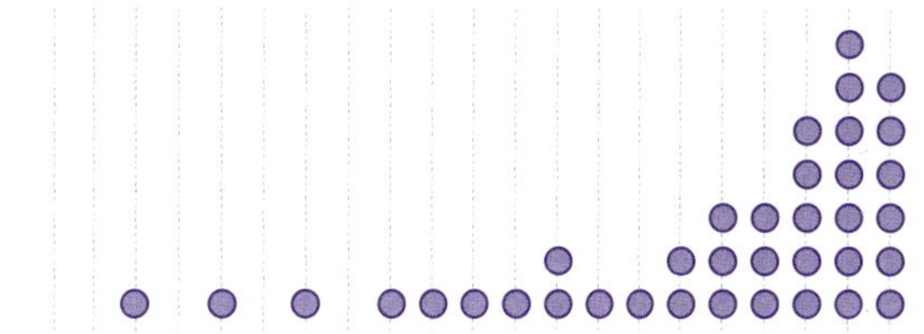

Bimodal
(there are two peaks)

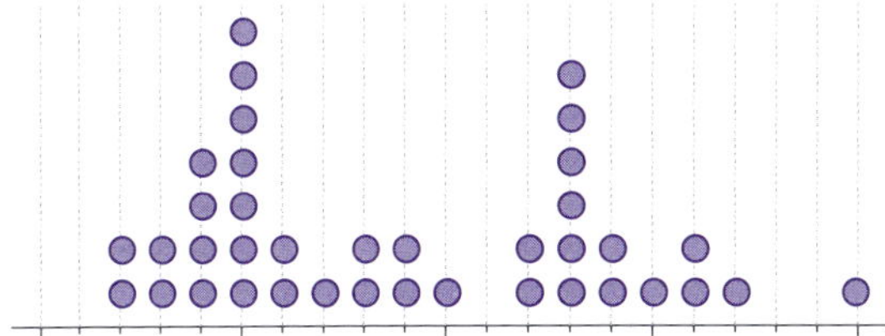

Uniform
(looks like a box)

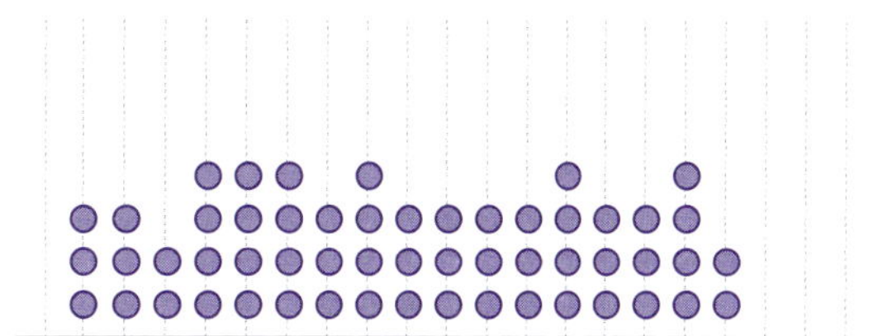

Triangular

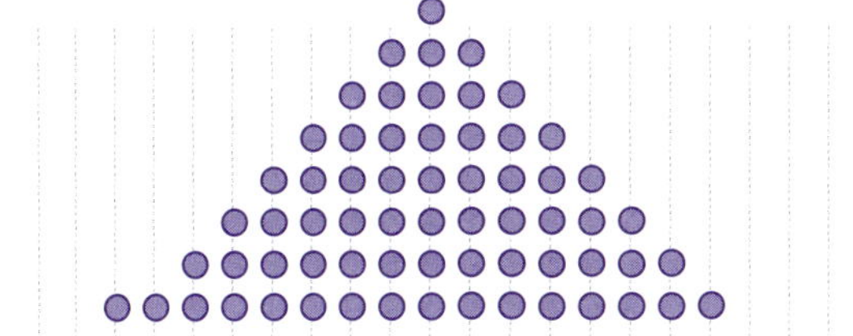

Irregular
(no real pattern)

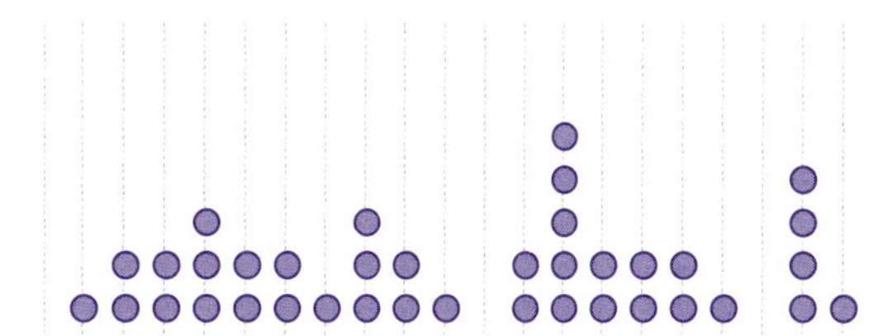

What shape do these graphs tend towards?

1

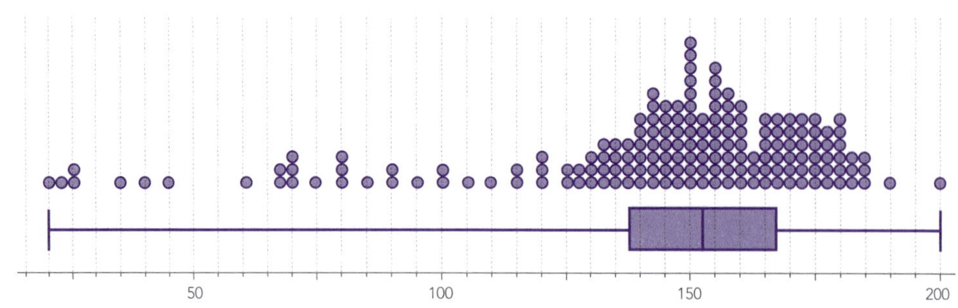

2

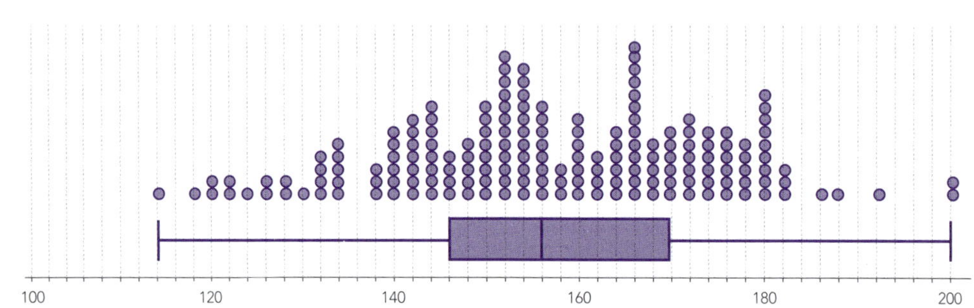

3

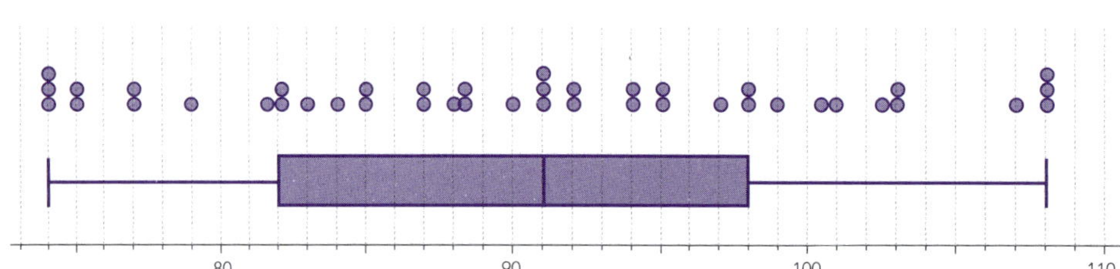

4

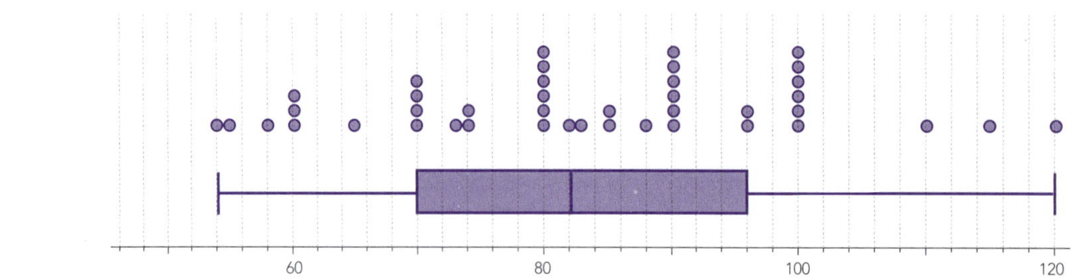

ISBN: 9780170370417

2 Comparing shapes

It is unlikely that your sample will look perfectly like the shapes on page 39. Therefore we can use language like 'tends towards …'.

The graphs below show the heights of some Year 10 students.

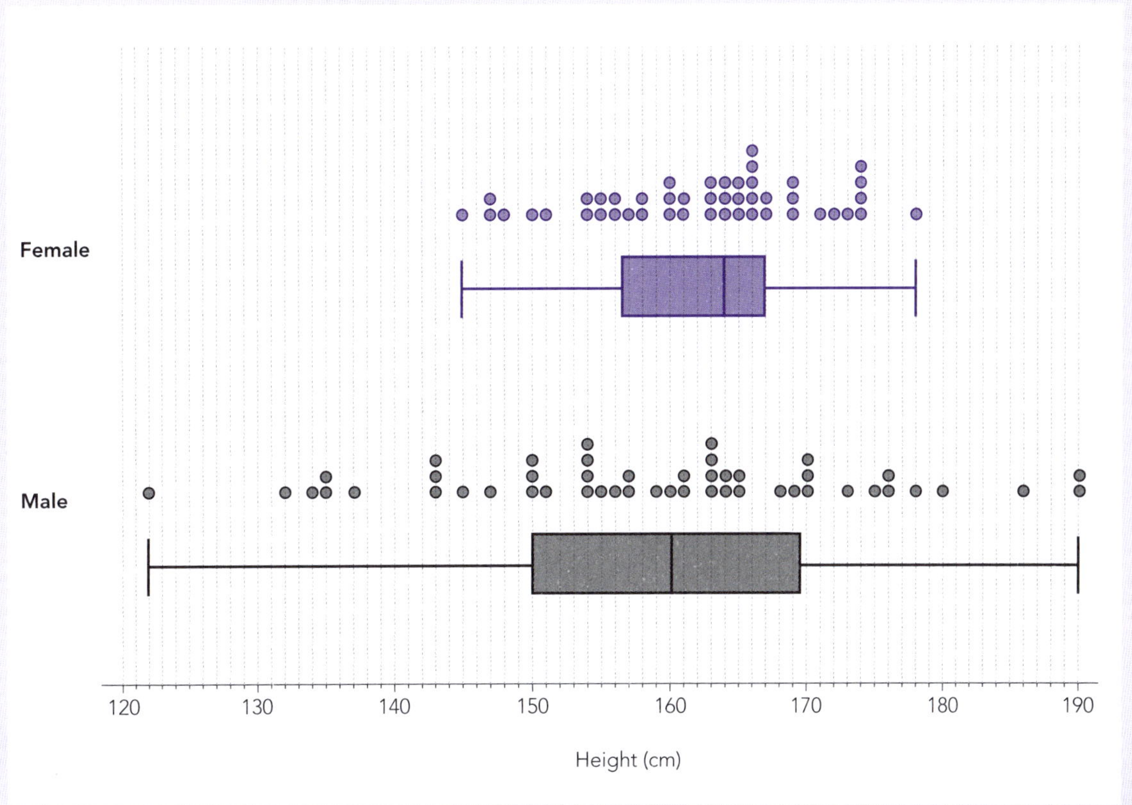

What do you see?
I notice that the females' distribution tends towards a normal distribution and the distribution of heights for males tends towards an irregular shape.

What does this mean for the samples?
This means that the boys' heights are more variable, whereas the girls' heights are more clustered around the median.

Why might this be the case?

If the samples were bigger, how would this change?
We would be more certain of the shapes and our conclusion.

ISBN: 9780170370417

Compare the shapes of the following samples of test marks for Year 11 students:

1

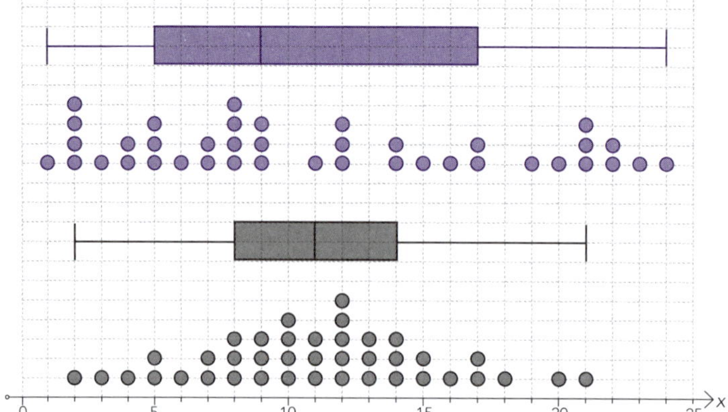

Boys

2

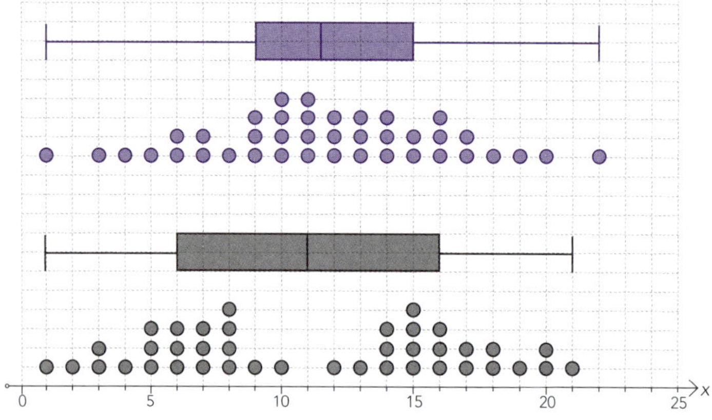

Boys

3

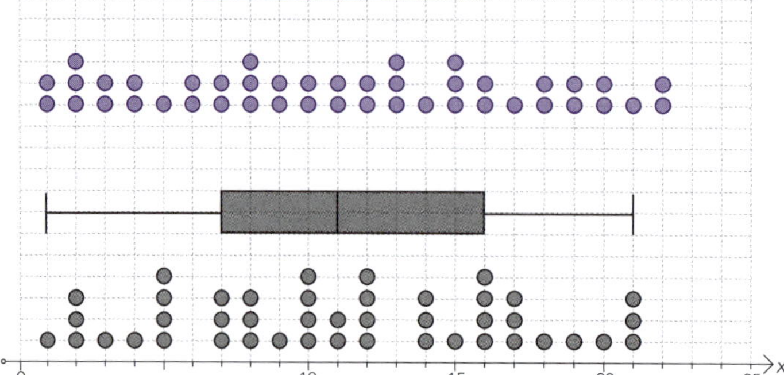

Boys

 ISBN: 9780170370417

Spreads

- **Size:** you need to compare the widths of the IQRs and the ranges.
- **Position:** you need to compare where the boxes lie — whether they are to the left or right of each other.
- **Overlap:** you need to draw an arrow where the boxes overlap, and discuss this.

Example 1: These box plots show History and Biology marks for a Year 11 class:

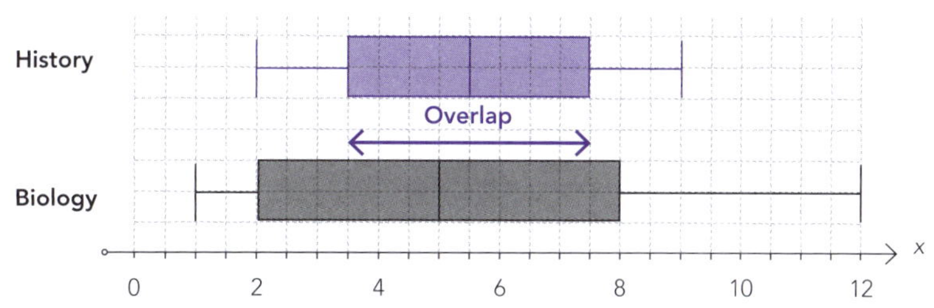

What do you see?
I notice that the IQR and the range for History marks are both smaller than those for Biology. I also notice that the middle 50% of History scores lie within the middle 50% of Biology scores.

What does this mean for the sample?
The smaller IQR for History means that there are more values clustered around the median, whereas the middle 50% of the Biology scores will be more spread out. Because the History box overlaps completely with the Biology box, marks in the two subjects were similar for this Year 11 class.
Note: You will have the statistics as well so you can be specific with numbers.

Example 2: These box plots show Science and English marks for a Year 11 class:

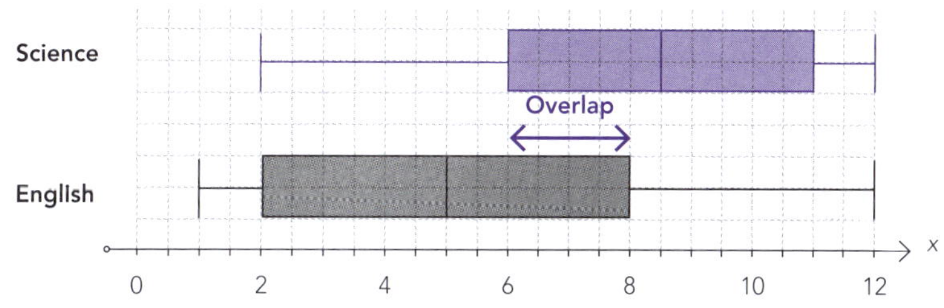

What do you see?
The box for Science is further to the right than the box for English.The IQR and the range for Science are slightly smaller than those for English. 75% of the Science marks are higher than the median score for English, so the data between the LQ and the median for the Science overlaps with most of the data between the median and the UQ for English.

What does this mean for the sample?
This means that the Science marks are slightly more spread out than the English marks. It also means that the majority of those in this class have scored better marks for Science than they have for English.

ISBN: 9780170370417

Compare the sizes, overlaps and spreads of the following samples of test marks for samples of Year 11 students.

1

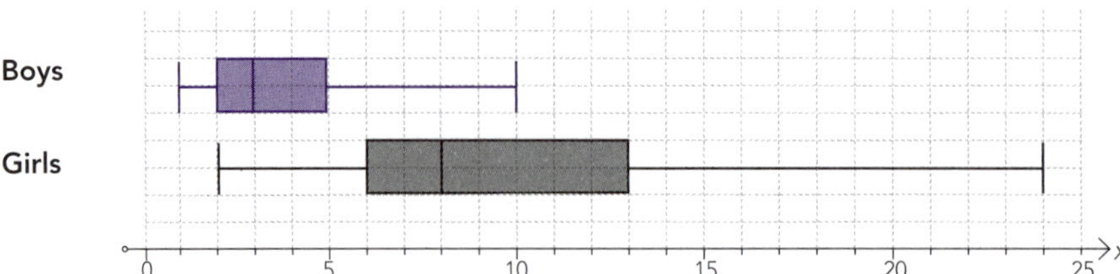

2

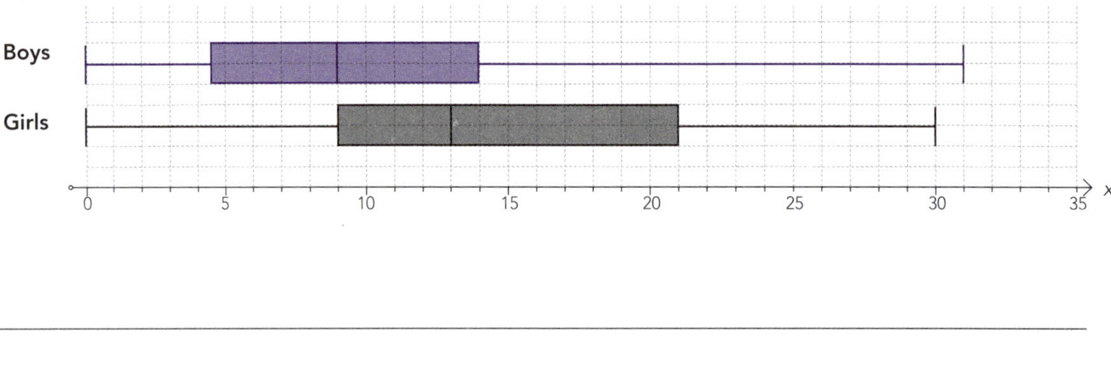

3

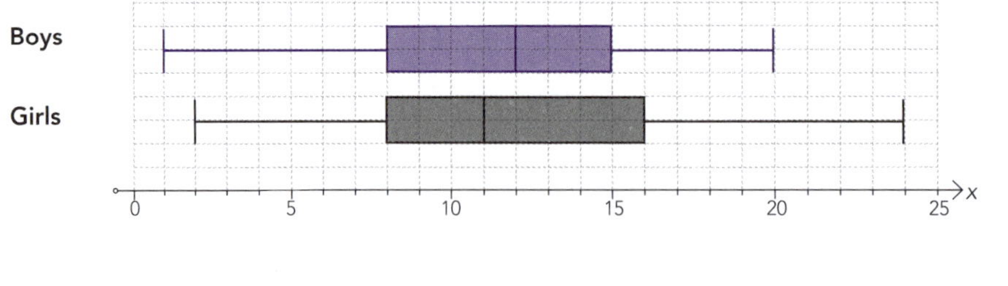

ISBN: 9780170370417

Unusual features

These could be an unusual point or a cluster.

These must be VERY obvious.
Don't comment on them if they are not there!

This shows the arm spans (cm) of a group of students.

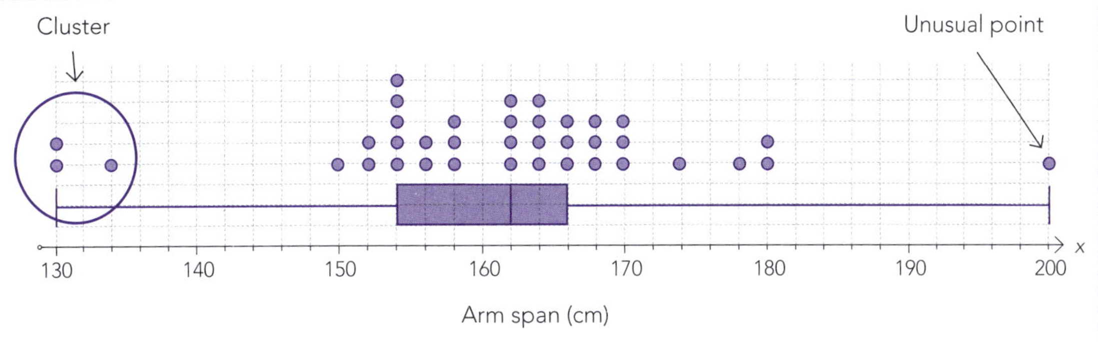

What do you see?
There is a cluster of smaller values and one larger value that are significant distances from most of the data.

What does this mean for the sample?
This means that there are three students with a much smaller arm span than the majority of the sampled students and there is one with a much larger arm span.

Why might we have these clusters or unusual features? Link to the context if possible.

Cleaning data

- This means removing pieces of data that are **obvious** mistakes.
- If possible, check the piece of data, re-measure the subject, check the units, etc.

Examples: **1** Heights of Year 11 students in cm:
156, 149, 181, 194, **276**, 187, 143, 173, …

Remove this piece.

2 Marks out of 20:
18, 14, 9, 11, 16, 13, **27**, 20, 16, 15, …

Remove this piece.

- **Always** state in your report whether or not you have removed pieces of data, and justify your decisions.

Do not remove unusual points unless you have a good reason for doing so!

ISBN: 9780170370417

Describe any unusual features in the following (not all graphs have them).

1

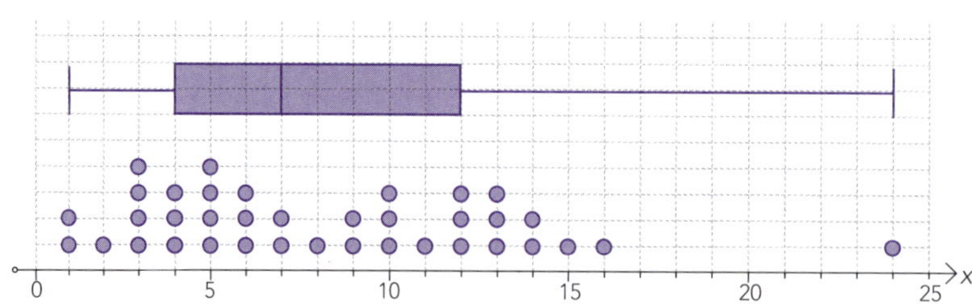

2

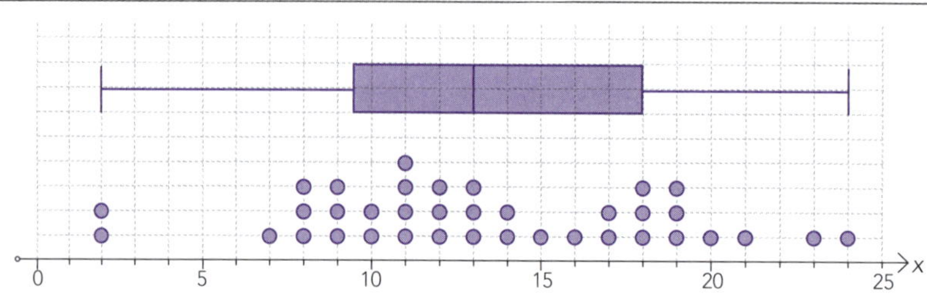

3

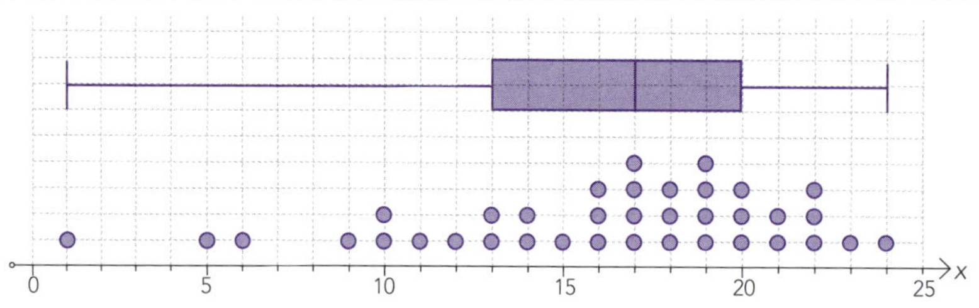

Cleaning data

State which points you would remove (if any) from the following data sets, and why.

1 Arm spans of Year 11 students (cm):

149, 168, 153, 166, 175, 171, 189, 171, 173, 160, 281, 144, 174, 168, 156, 153

2 Arm spans of Year 11 students (cm):

143, 144, 158, 193, 161, 1.85, 141, 177, 161, 154, 168, 181, 179, 168, 156, 159

3 Arm spans of Year 11 students (cm):

167, 151, 183, 211, 159, 172, 162, 183, 161, 166, 180, 148, 169, 168, 147, 175

4 Arm spans of Year 11 students (to the nearest cm):

159, 171, 150, 181, 177, 160, 189.5, 169, 163, 170, 181, 148, 166, 170, 191, 157

  ISBN: 9780170370417

Putting it together

When you are writing comments, ensure that you compare both groups for every aspect of your report.

Example:
This shows shoulder widths for a sample of high school students at Paradise High School.

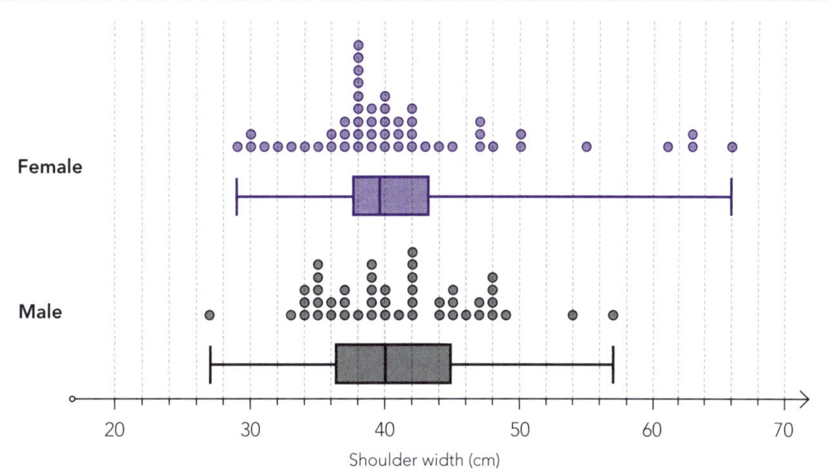

	Min.	LQ	Median	Mean	UQ	Max.	Sample size
Female	29	37.75	39.5	41.44	43.25	66	52
Male	27	36.25	40	40.89	45	57	46

Centres

The median shoulder width for males is 0.5 cm greater than that for females. The mean shoulder widths are the other way around, probably because of the tail of larger values for females. These would increase the mean but not the median. There is not a lot of difference in the positions of the boxes.

This means that, for these samples, there is not a lot of difference between the shoulder widths of males and females.

Shapes

The shape of the distribution of female shoulder widths tends to a normal distribution, but with a slight skew to the right. The shape for the boys is irregular.

Spreads

The IQR for male shoulder widths (8.75 cm) is greater than that for female shoulder widths (5.5 cm). This means that for these samples, the male shoulder widths were more spread than the female shoulder widths.

The boxes are in similar positions, but the box for female shoulder widths lies within that for males, so it totally overlaps the box for male shoulder widths. This supports the conclusions that, for these samples, the boys' shoulder widths are more spread than the girls', and that the boys' shoulder widths are only a little bigger than the girls' shoulder widths.

Unusual features

For the female shoulder widths, there is a group of four girls who have much broader shoulders than any of the boys. The broadest shoulders for the boys is 57 cm, but all four girls have shoulders broader than 60 cm. It would be interesting to see if these girls had anything in common, e.g. are they all swimmers?

For the males, there is one unusual value: one boy has very narrow shoulders of 27 cm, which is less than the narrowest shoulder width for any girl (29 cm).

I chose not to remove any of these values, as I felt that they were all feasible.

Discuss the centres, shapes, spreads and unusual features for the following samples.

1 The following shows the arm spans of 51 female and 47 male Year 11 students at Paradise High School.

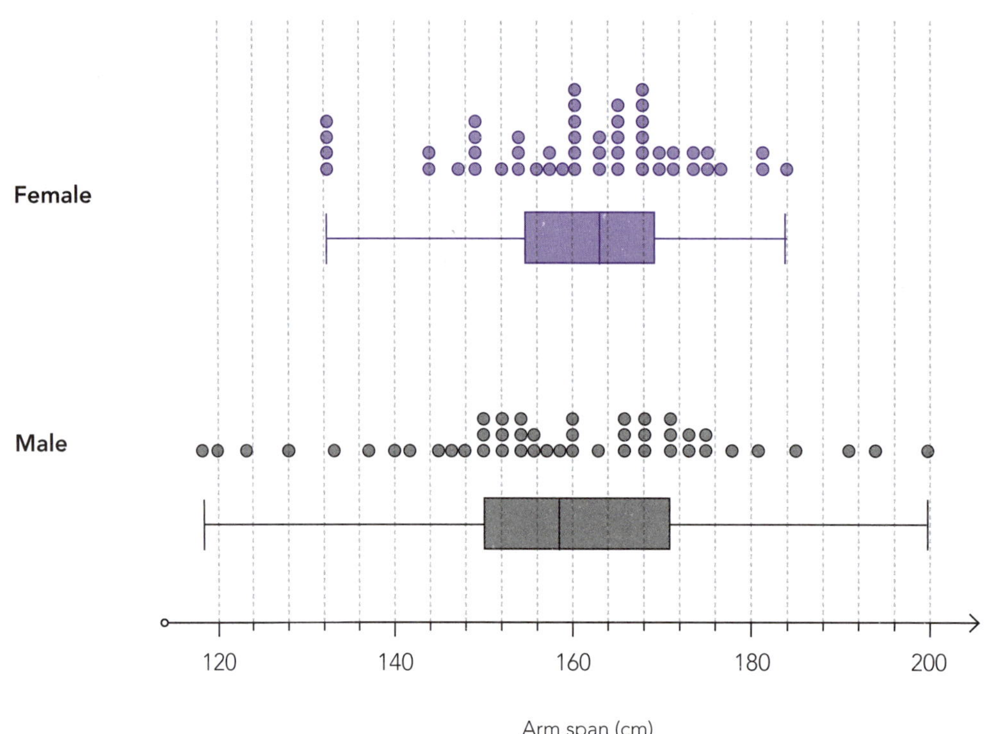

	Min.	LQ	Median	Mean	UQ	Max.	Sample size
Female	132	154.5	163	160.9	169	184	51
Male	118	149	159	158.1	171	200	47

Centres

ISBN: 9780170370417

Shapes

Spreads

Unusual features

ISBN: 9780170370417

2 The following shows the lengths of the right foot (cm) of 47 female and 45 male Year 11 students at Paradise High School.

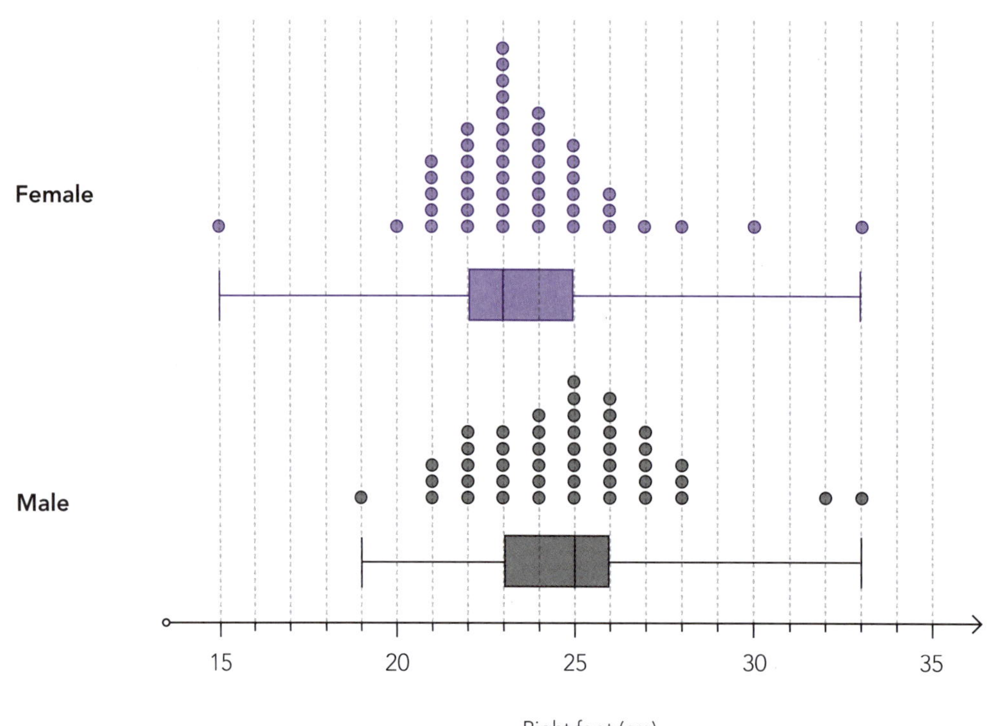

	Min.	LQ	Median	Mean	UQ	Max.	Sample size
Female	15	22	23	23.57	25	33	47
Male	19	23	25	24.82	26	33	45

Centres

ISBN: 9780170370417

Shapes

Spreads

Unusual features

ISBN: 9780170370417

3 The following shows the numbers of texts sent in a week by 56 female and 44 male Year 11 students at Paradise High School.

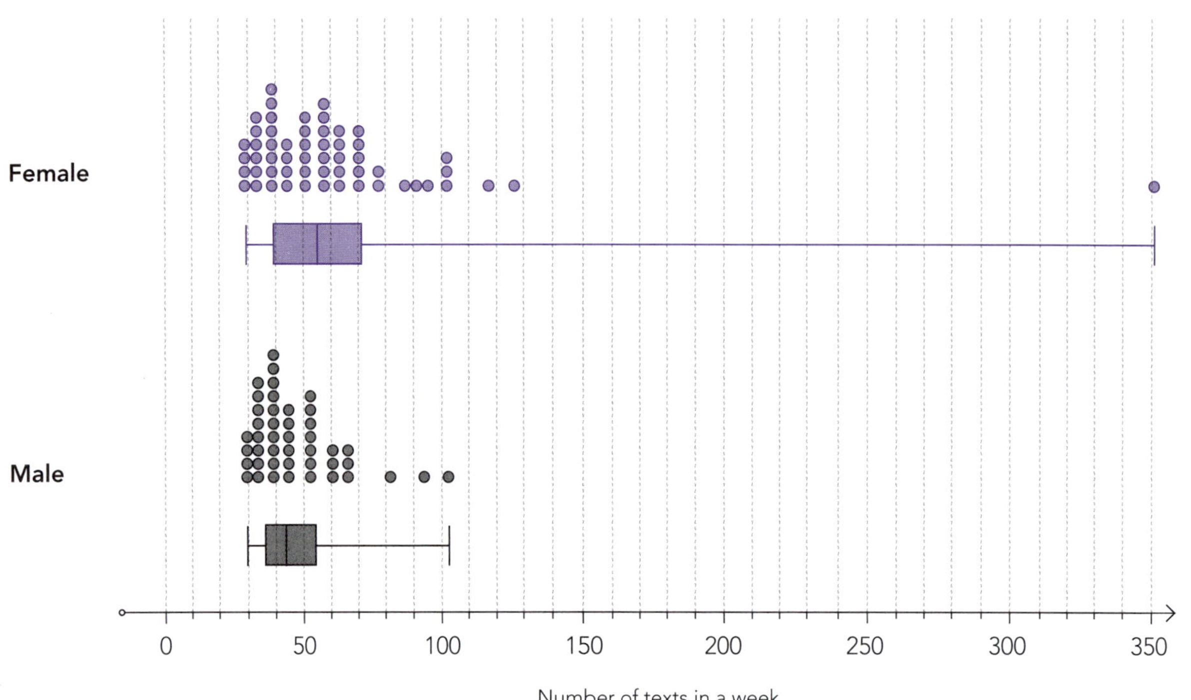

	Min.	LQ	Median	Mean	UQ	Max.	Sample size
Female	28	38	54	61.98	70.5	352	56
Male	29	35.5	42.5	47.48	53.5	101	44

ISBN: 9780170370417

4 The following shows the shoulder widths (cm) of 48 Year 12 and 48 Year 9 students at Paradise High School.

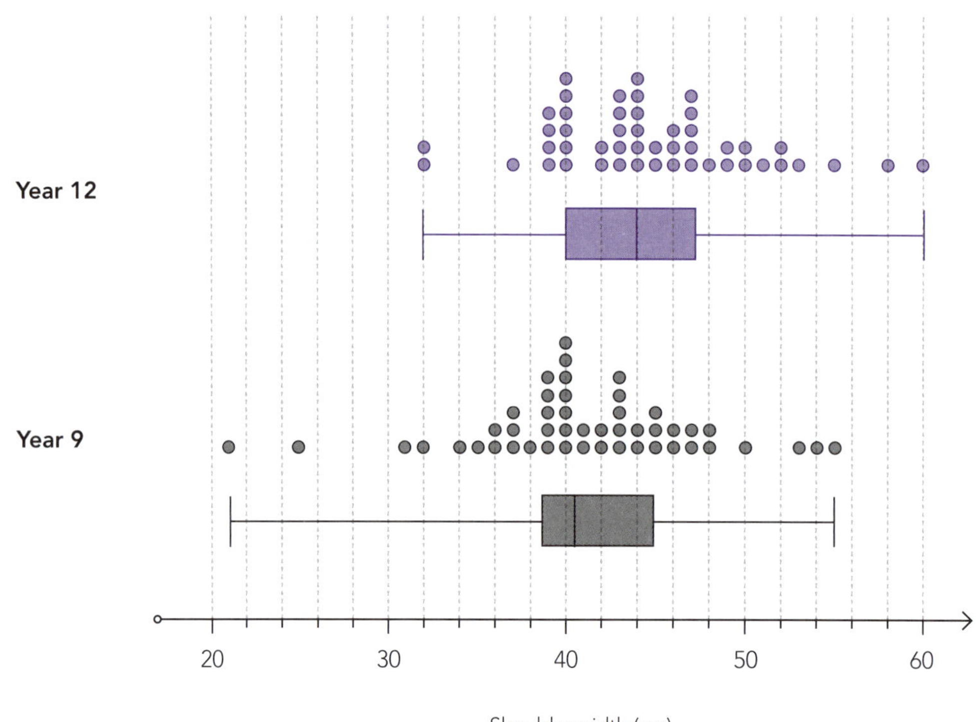

	Min.	LQ	Median	Mean	UQ	Max.	Sample size
Year 12	32	40	44	44.79	47.5	60	48
Year 9	21	38.5	40.5	41.04	45	55	48

ISBN: 9780170370417

Making the call about the population

The purpose of this analysis is to answer your question and make a call about which group is larger or smaller in the population.

Method 1

- We can decide whether there is a difference between **samples** by comparing their box plots.
- We can also decide whether there might be a difference between **populations** by comparing their box plots. We need to consider the positions of the medians and middle 50% of both samples.

The box plots below show the lengths of lizards in samples from populations A and B.

Case 1: The boxes of each sample overlap with each other, and both medians lie within the box of the other sample.

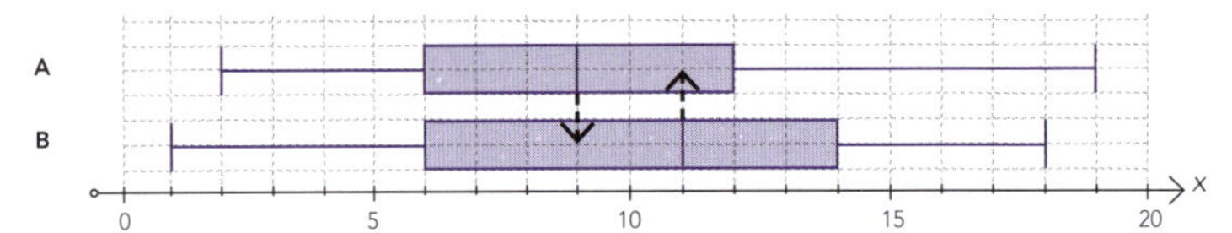

We **cannot** conclude that individuals in population B tend to be longer than those in population A, because both of the medians lie within the middle 50% of the other sample, and the boxes overlap.

Case 2: The boxes of each sample overlap with each other, and one or both medians lie outside the box of the other sample.

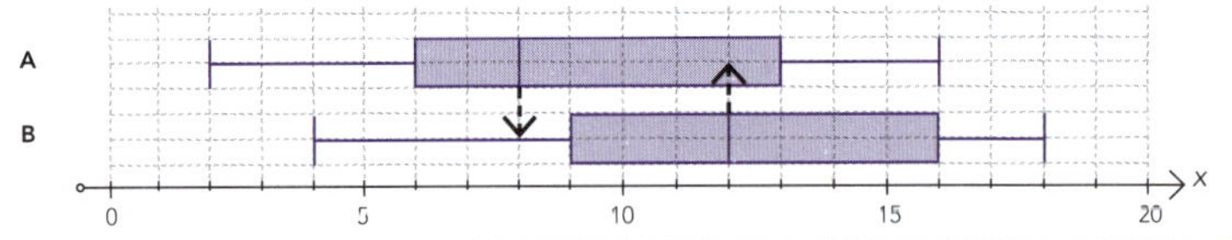

We **can** conclude that individuals in population B **are likely**, on average, to be longer than those in population A because, although the middle 50% of each sample overlap, one median lies outside the box of the other group.

Case 3: The boxes of each sample do not overlap with each other.

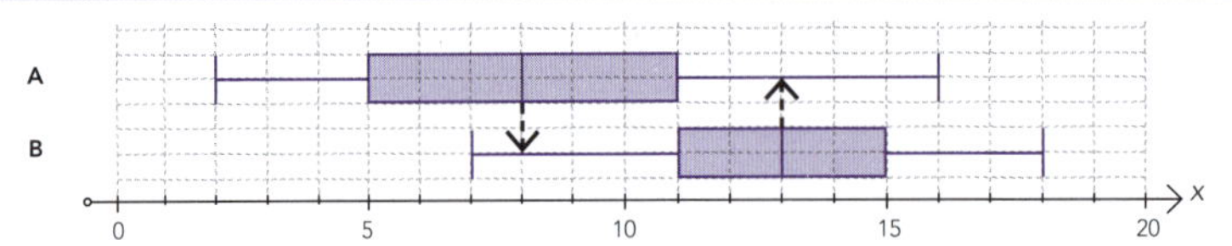

We **can** conclude that individuals in population B **will** tend to be longer on average than those in population A because the boxes do not overlap and both medians lie outside the other box.

ISBN: 9780170370417

Write conclusions about the lengths of lizards in the populations for each of the pairs of sample box plots:

1

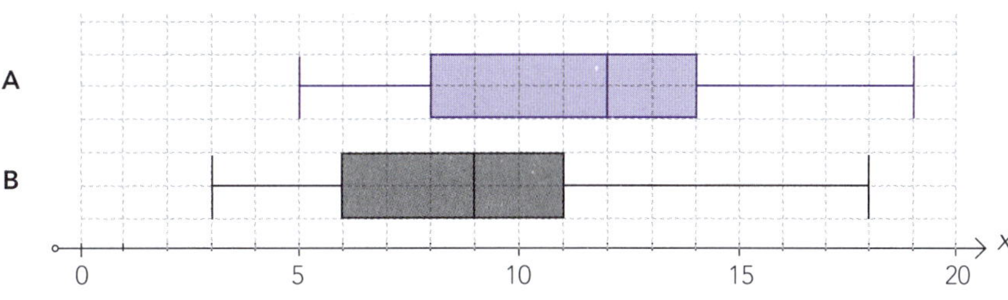

2

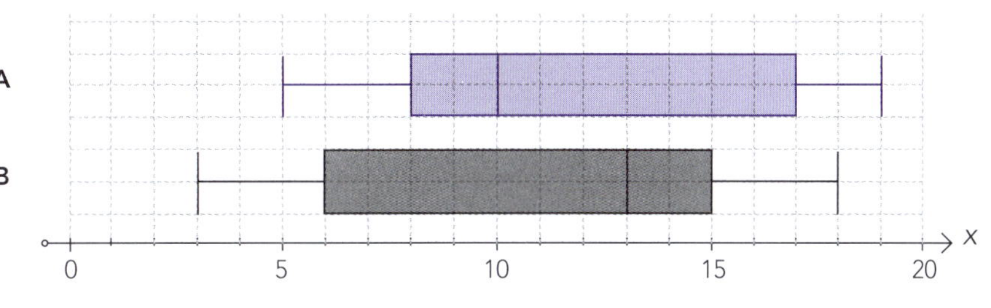

3

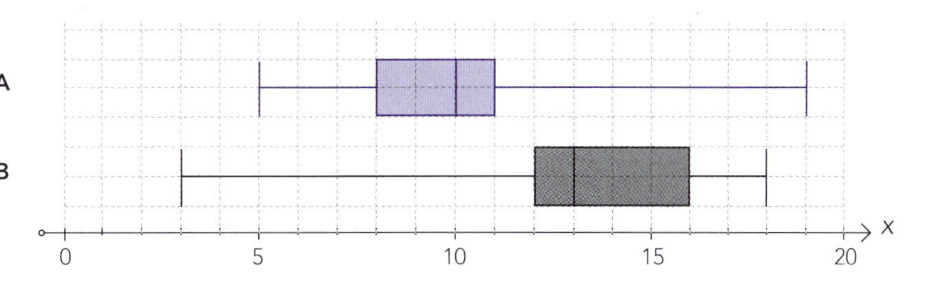

4

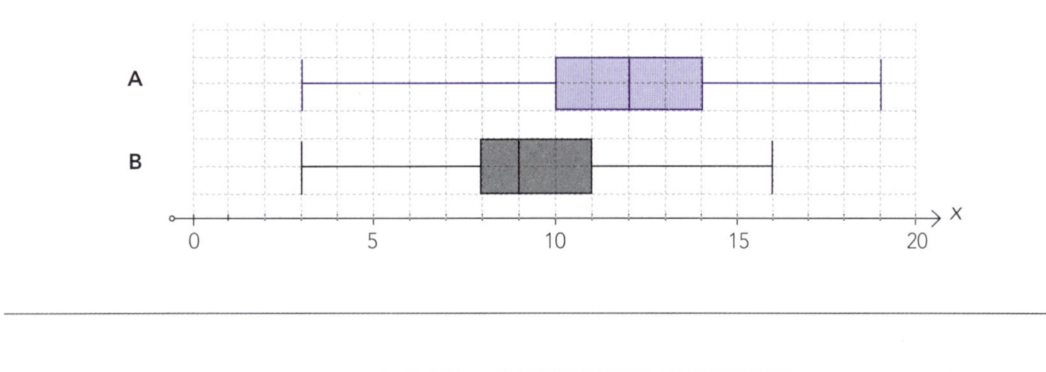

ISBN: 9780170370417

5

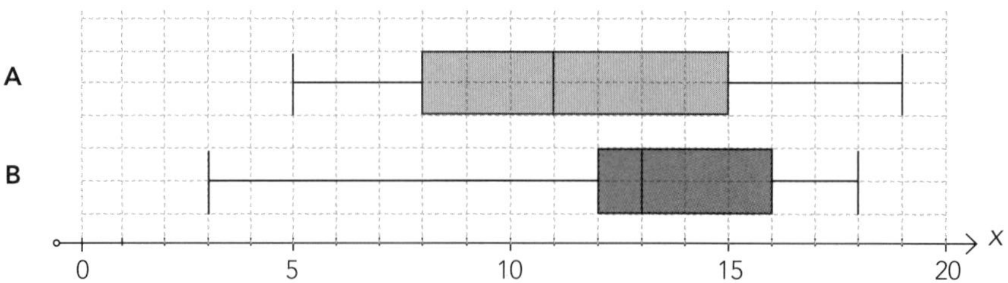

6

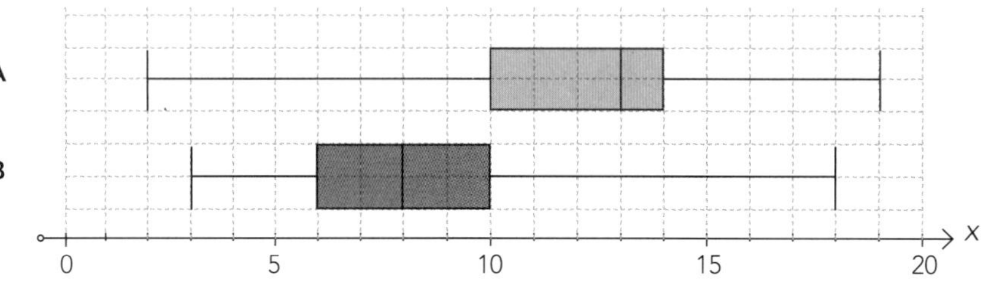

7

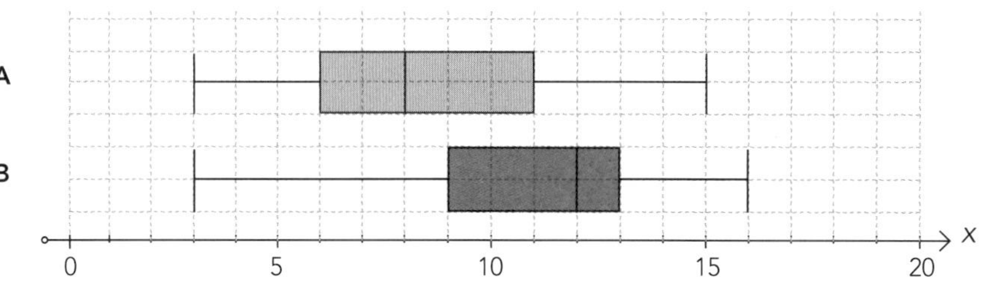

8

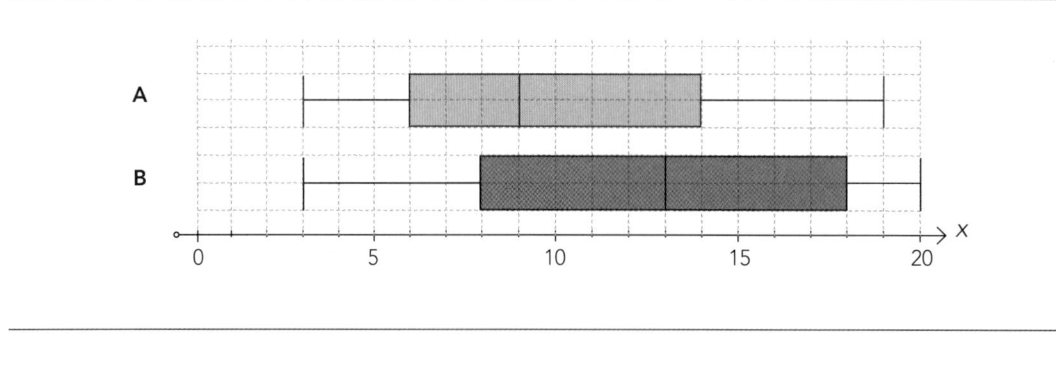

ISBN: 9780170370417

Method 2

Another way to make the call is to calculate the difference between the medians and compare it to the overall visual spread.

OVS = **O**verall **V**isible **S**pread

DBM = **D**ifference **B**etween **Medians**

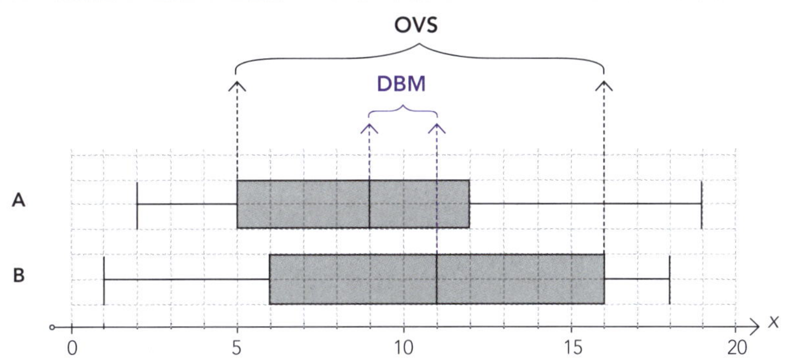

For a sample size of 30:
The **DBM** is more than **⅓ of the OVS** ⇒ individuals in population B tend to be bigger those in population A.

For a sample size of 100:
The **DBM** is more than **⅕ of the OVS** ⇒ individuals in population B tend to be bigger than those in population A.

Method for sample size of 30:

Step 1: Draw lines to show the **OVS**:

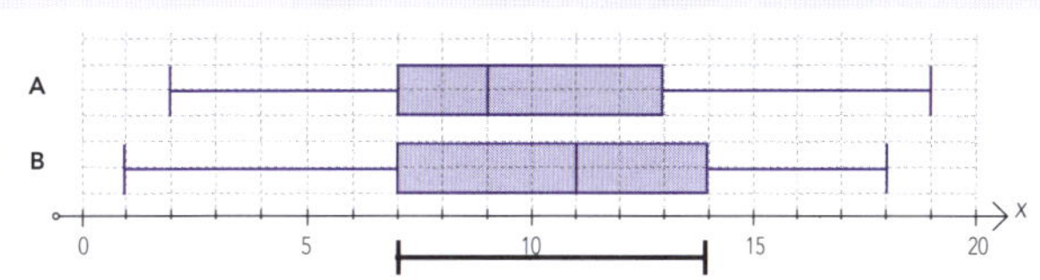

Step 2: By eye, divide the **OVS** into ⅓s:

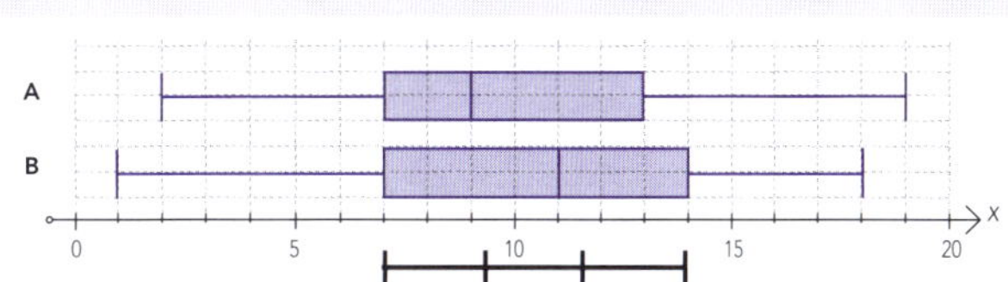

Step 3: Draw lines for the **DBM**:

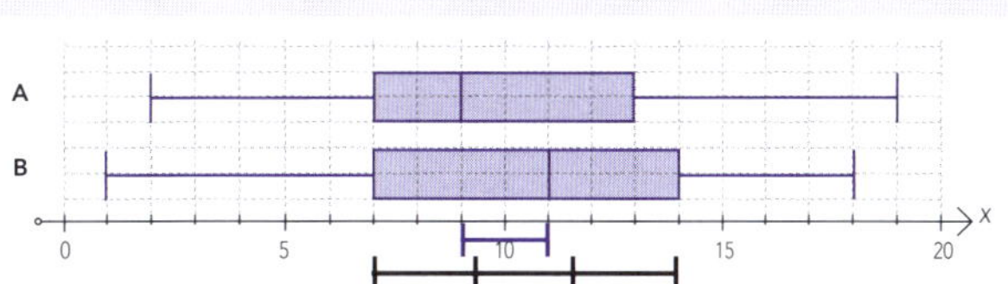

Step 4: Compare the **DBM** with the $\frac{\textbf{OVS}}{\textbf{3}}$.

*Because the **DBM** is **less** than the $\frac{\textbf{OVS}}{\textbf{3}}$, we **cannot** conclude that individuals in population B will tend to be longer than those in population A.*

Note: We came to the same conclusion using Method 1.

ISBN: 9780170370417

Method for sample size of 100:

Step 1: Draw lines to show the **OVS**:

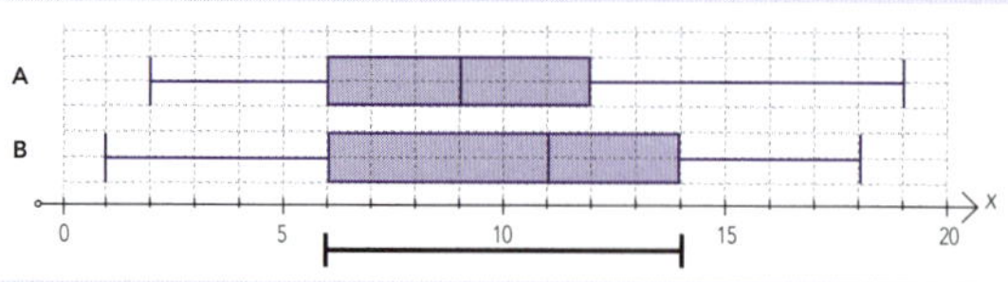

Step 2: By eye, divide the **OVS** into ⅕s:

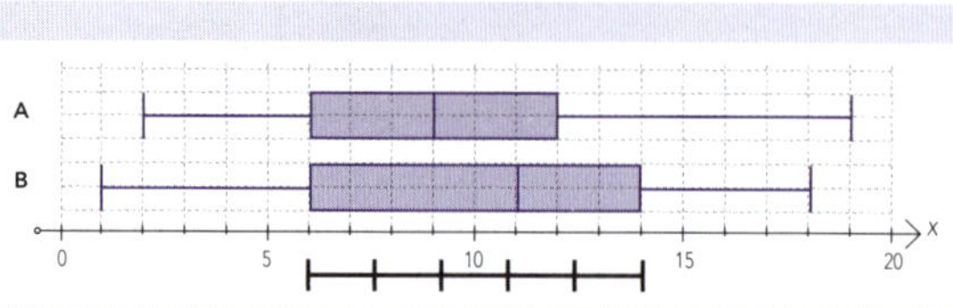

Step 3: Draw lines for the **DBM**:

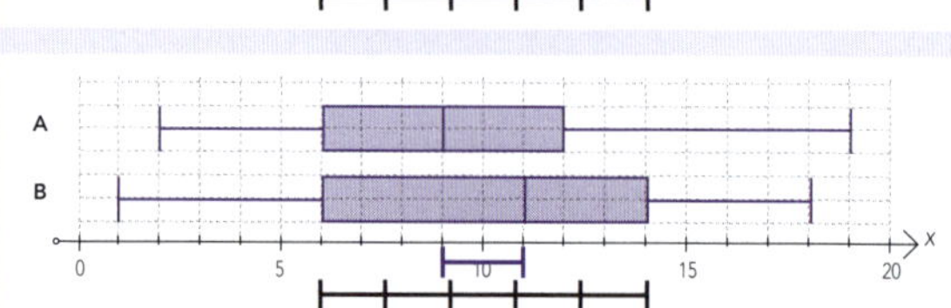

Step 4: Compare the **DBM** with the $\frac{\textbf{OVS}}{\textbf{5}}$.

*Because the **DBM** is **more** than the $\frac{\textbf{OVS}}{\textbf{5}}$, we can conclude that individuals in population B **will** tend to be longer than those in population A.*

Note: We have not come to the same conclusion as we did with Method 1 because the sample is much bigger.

In general: Do **not** measure or calculate these intervals. If you cannot tell that one is bigger by eye, then do **not** conclude that there is a difference in the population.

1 Sample size of 30

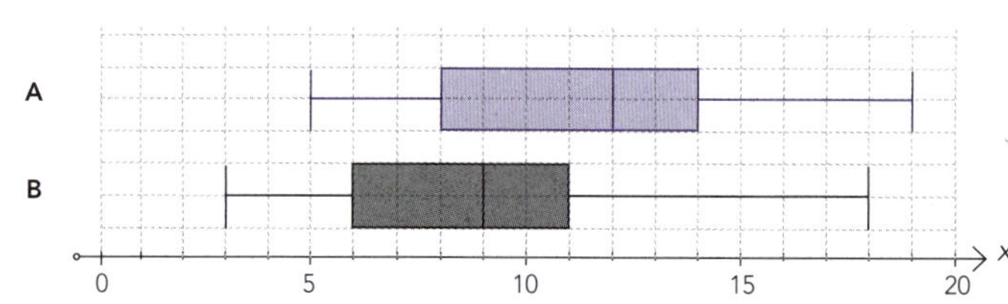

Compare your conclusion with that for number **1** in the previous exercise: is it the same?

2 Sample size of 100

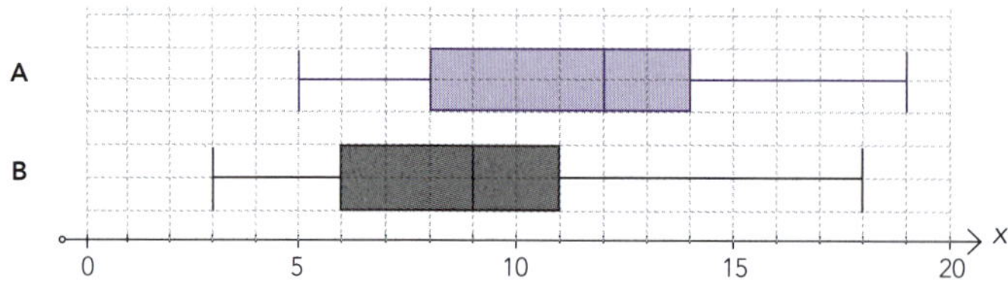

Compare your conclusion with that for number **1** in the previous exercise: is it the same?

 ISBN: 9780170370417

3 Sample size of 30

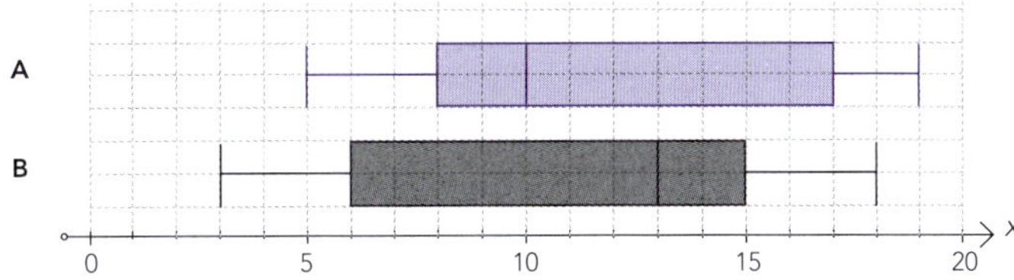

Compare your conclusion with that for number **2** in the previous exercise: is it the same?

4 Sample size of 100

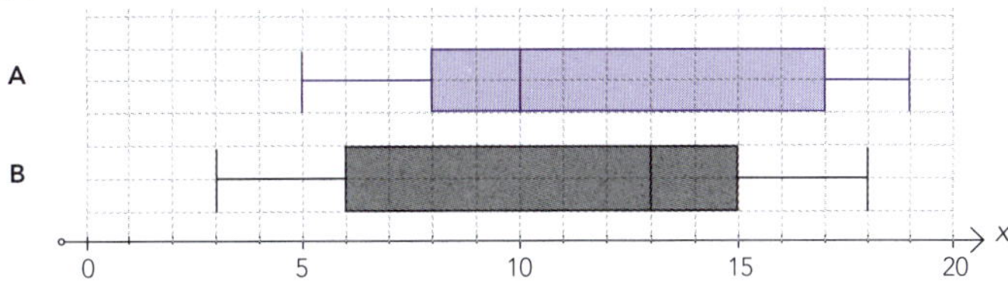

Compare your conclusion with that for number **2** in the previous exercise: is it the same?

5 Sample size of 30

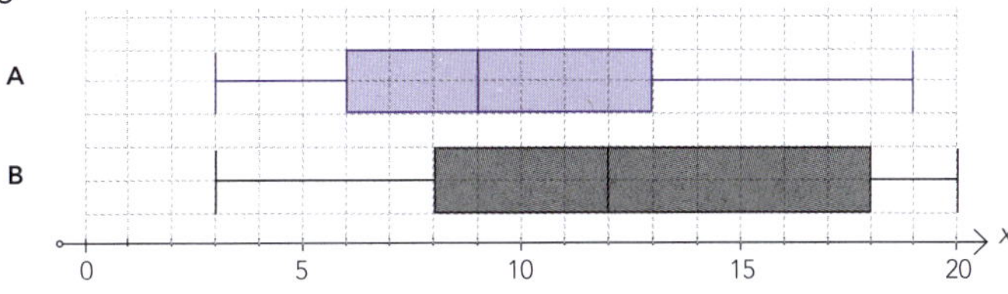

Compare your conclusion with that for number **8** in the previous exercise: is it the same?

6 Sample size of 100

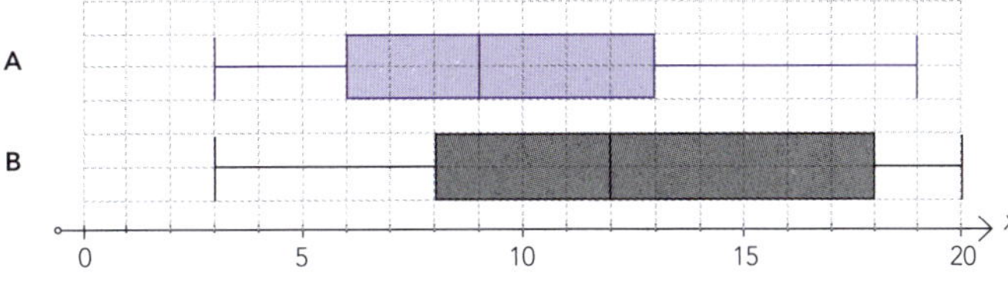

Compare your conclusion with that for number **8** in the previous exercise: is it the same?

ISBN: 9780170370417

Questions and conclusions about the population

Write a suitable question, and write a conclusion regarding the population for each of the following:

1 The following shows the heights of 50 males and 50 females from Year 13 at Paradise High School.

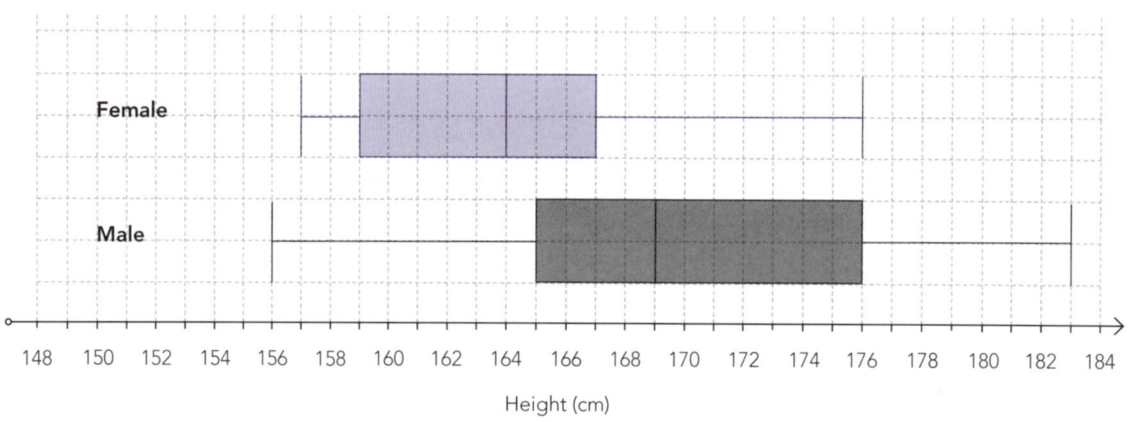

Question

Conclusion

ISBN: 9780170370417

2 The following show the hourly wages for part-time jobs of 50 males and 50 females from Year 13 at Paradise High School.

Question

Conclusion

ISBN: 9780170370417

3 The following shows the numbers of texts sent in a week by 50 males and 50 females from Year 13 at Paradise High School.

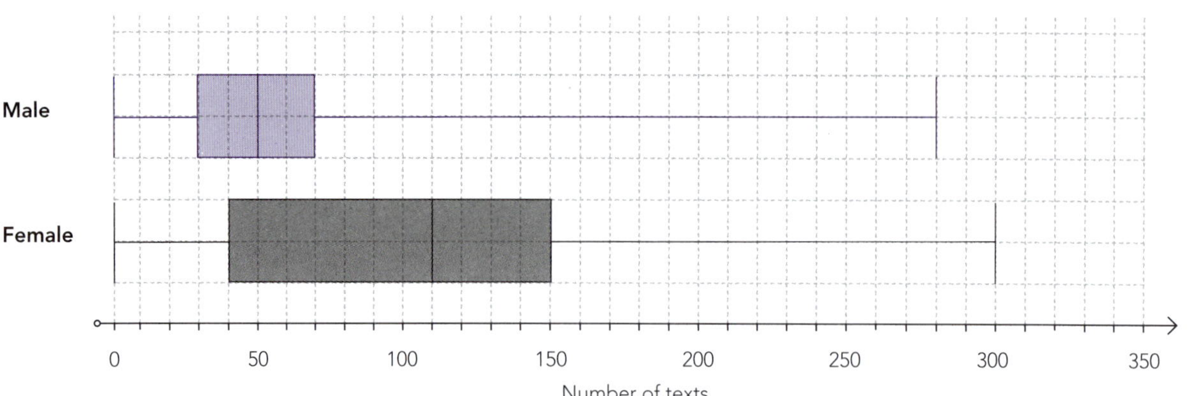

Question

Conclusion

ISBN: 9780170370417

Reflections

These are some reflective questions that you could consider answering in your report.

- Was my sample big enough?
- Was there any bias in my samples?
- Was my sample representative of the population?
- Was the source of my data reliable?
- Did I clean my sample data?
- How confident am I of my conclusions?
- Do my conclusions seem reasonable?
- Are all my statements true?
- Did I answer my question?

ISBN: 9780170370417

Writing about data

You need to be careful how you write about data.

Points

- The **mean** is a 'point'.
- These are also the **five** 'points' on a box plot.

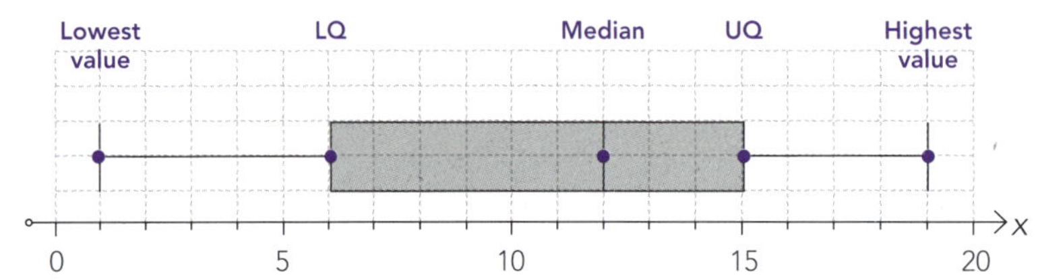

Both of these are **points**, so either they are the same or one is bigger than the other.

Examples:

✗ 'The lower quartile of A overlaps with the upper quartile of B.'

✓ 'The lower quartile of A is the same as the upper quartile of B. This means that, for these samples, 75% of the values for A are bigger than the upper quartile for B.'

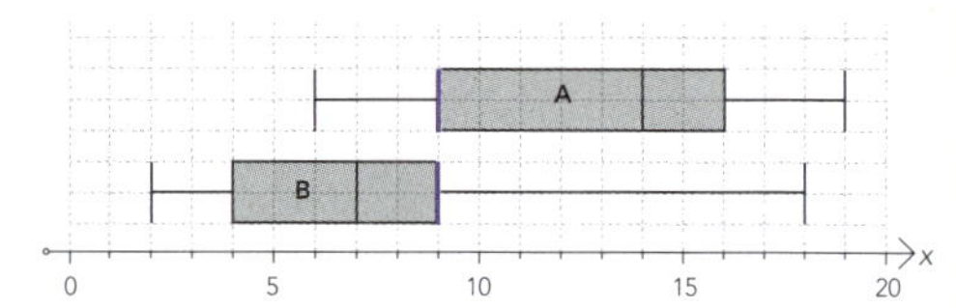

Interval widths

There are two interval widths:

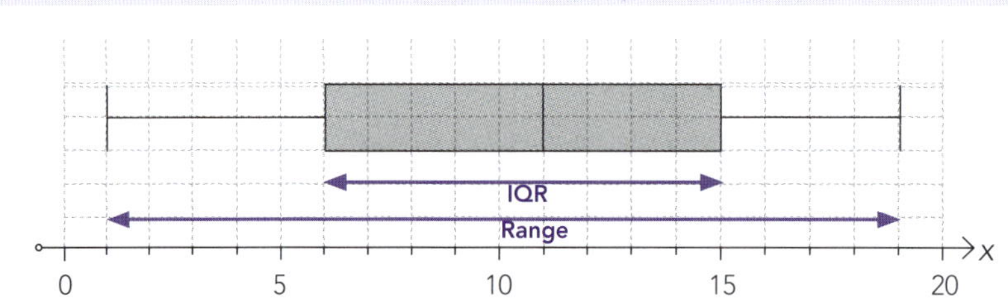

- In this case, the IQR = 9 and the Range = 18.
- On their own, they tell you **nothing** about **where** the data lies: they only tell you how widely it is **spread**.

The IQR is a **width**, not a position, so IQRs can be the same or one can be wider than the other.

Examples:

✗ 'The IQR of A is further to the right than the IQR of B.'

✓ 'The IQR of A is bigger than the IQR of B. This means that, for these samples, the middle 50% of the data (or the box) for A is more widely spread than the middle 50% of the data (or the box) for B.'

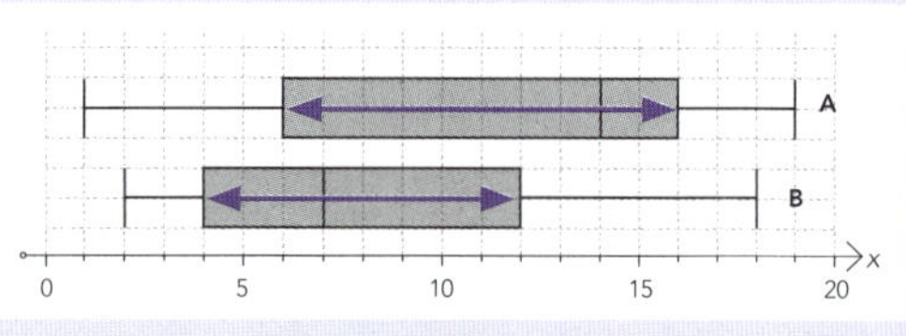

ISBN: 9780170370417

✓ 'The middle 50% of the data (or the box) for A is further to the right than the middle 50% of the data (or the box) for B. This means that, for these samples, the middle 50% of the …….. in A tend to be bigger/longer, etc. than the middle 50% of …….. in B.'

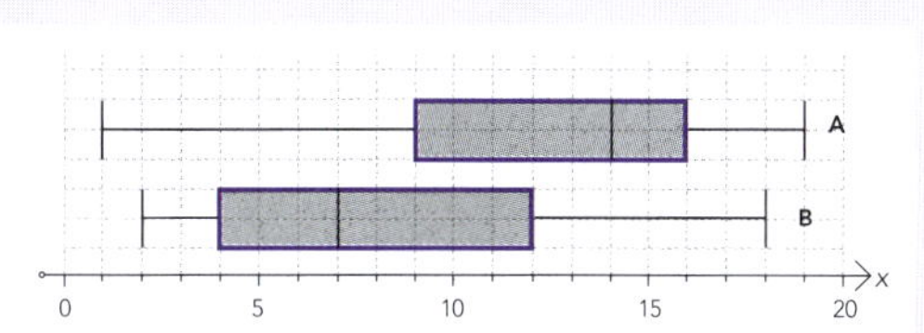

Intervals

- Intervals have a fixed **location and width**.
- Use terms such as these:

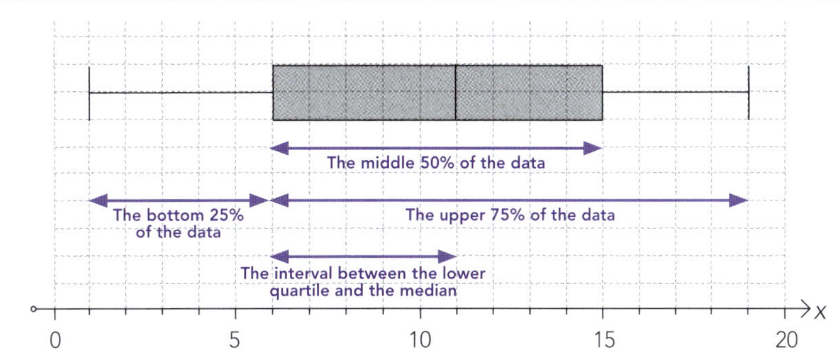

- In this case, the bottom 25% of the data is between 1 and 6, and it is 5 units wide.

✗ 'The middle 50% of A is bigger than the middle 50% of B.'

This is ambiguous. It might mean the middle 50% of A **is wider than** the middle 50% of B, or it might mean that the middle 50% of A lies **further to the right** than the middle 50% of B.

✓ 'The middle 50% of A is wider than the middle 50% of B. This means that, for these samples, the data in the middle 50% of A is more widely spread than that in the middle 50% of B.'

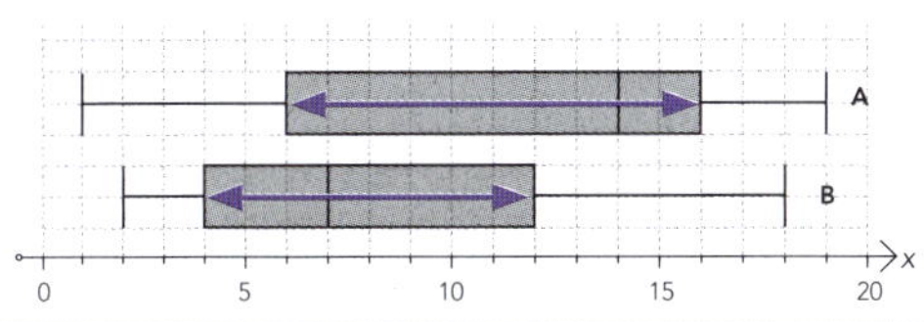

✓ 'The middle 50% of A lies to the right of the middle 50% of B. This means that, for these samples, 75% of the …….. in A tend to be bigger/longer, etc. than the UQ for B.'

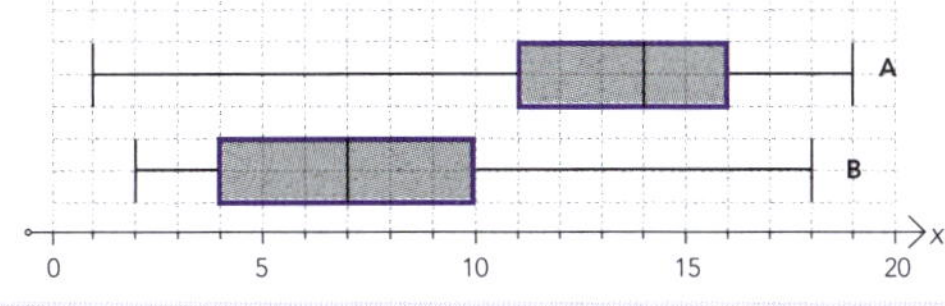

Be very careful in your report how you write about data.

ISBN: 9780170370417

Which box plot(s) **best** fit the following descriptions or statements? There may be one or several answers.

1 The box plot(s) with the data between the LQ and the median lying between 3 and 6.

____

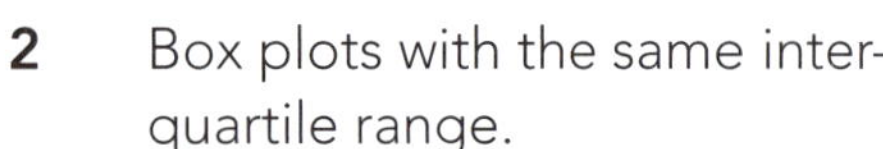

2 Box plots with the same inter-quartile range.

____

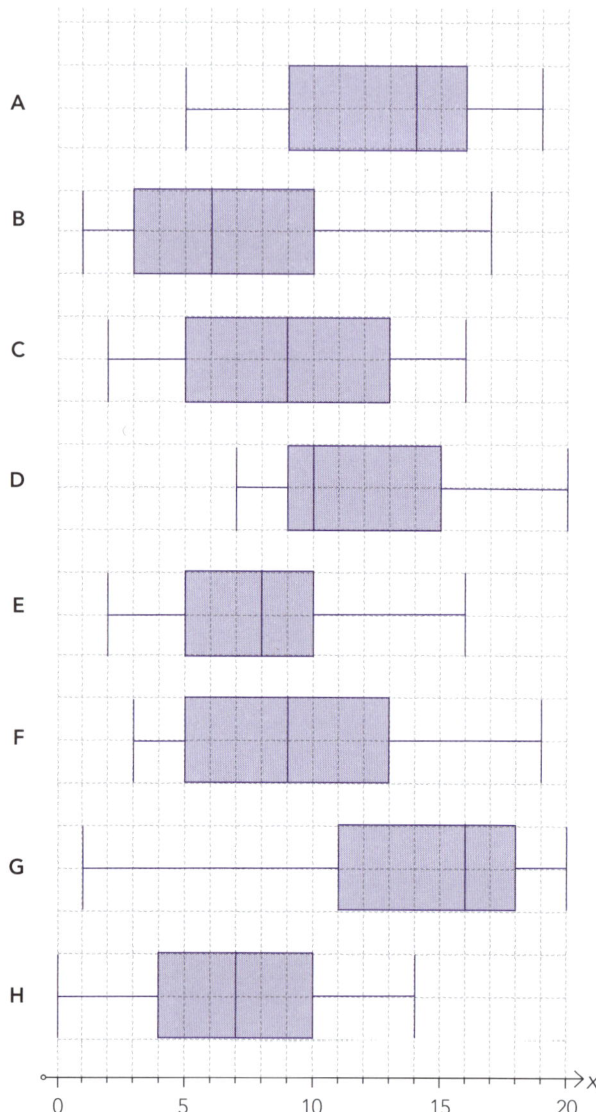

3 Pair(s) of box plots where the middle 50% of data of one does not overlap with the middle 50% of the other.

____

4 Pair(s) of box plots where the middle 50% of the data lies within the interval between the minimum and the LQ of the other.

____

5 Pair(s) of box plots where at least 75% of the data lies to the left of the lower quartile of the other.

____

6 Pair(s) of box plots where the interval between the minimum and the LQ overlaps with the interval between the UQ and the maximum.

____

7 The box plot(s) where the mean is most likely to be the same as the median.

____

8 Pair(s) of box plots where the interval between the LQ and the median of one lies below the lowest value of the other.

____

 ISBN: 9780170370417

State whether the following statements are true, false, ambiguous, or don't make sense. If they are false, ambiguous, or don't make sense, explain why.

9 The median of H is smaller than the IQR of D. This means that, in these samples, individuals in D tend to be bigger than individuals in H.

____

____

10 The interval between the LQ and the median of D lies within the interval between the median and the UQ of E. This means that, in these samples, individuals in E tend to be bigger than individuals in D.

____

____

11 The IQR of A is bigger than the IQR of B. This means that, in these samples, individuals in A tend to be bigger than individuals in B.

____

____

12 75% of the data for D lies above the median for C. This means that, in these samples, individuals in D tend to be bigger than individuals in C.

____

____

13 Box plot G is skewed to the right.

____

____

14 The interval between the median and the UQ of D is bigger than the interval between the median and the UQ of E. This means that the data in this interval is more spread out for sample D than it is for sample E.

____

____

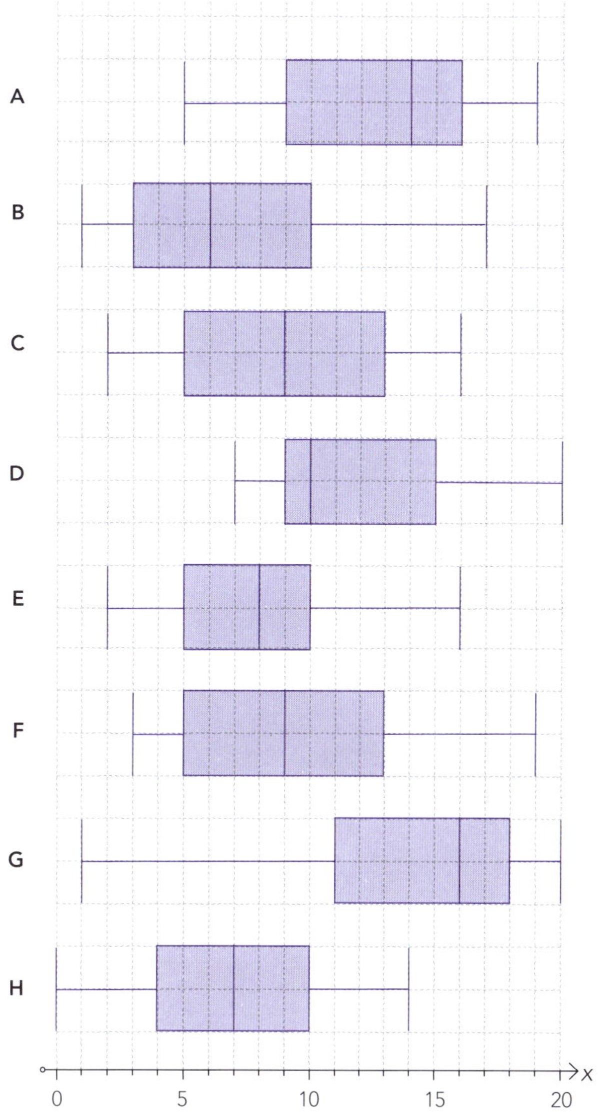

A
B
C
D
E
F
G
H
0 5 10 15 20 x

ISBN: 9780170370417

15 The following are examples of common mistakes made by students. Describe what is wrong with each.

Common mistakes	What is wrong with them?
The medians overlap.	
There is an unusual point so I will ignore/remove it.	
The boys' median is bigger than the girls' so there is a difference in the population.	
If I did another sample, I would get the same results.	
I have 30 in each group, so my conclusion is right.	
The UQ of the girls is bigger than the boys. This means the shape is irregular.	
The Year 9 inter-quartile range is to the right of the Year 11 median.	
There is a tail to the right, therefore the shape is a left skew.	

ISBN: 9780170370417

Practice tasks

Practice task one

The data below shows the lengths of little fingers of 58 Year 7 girls and 58 Year 7 boys. The population is New Zealand Year 7 students.

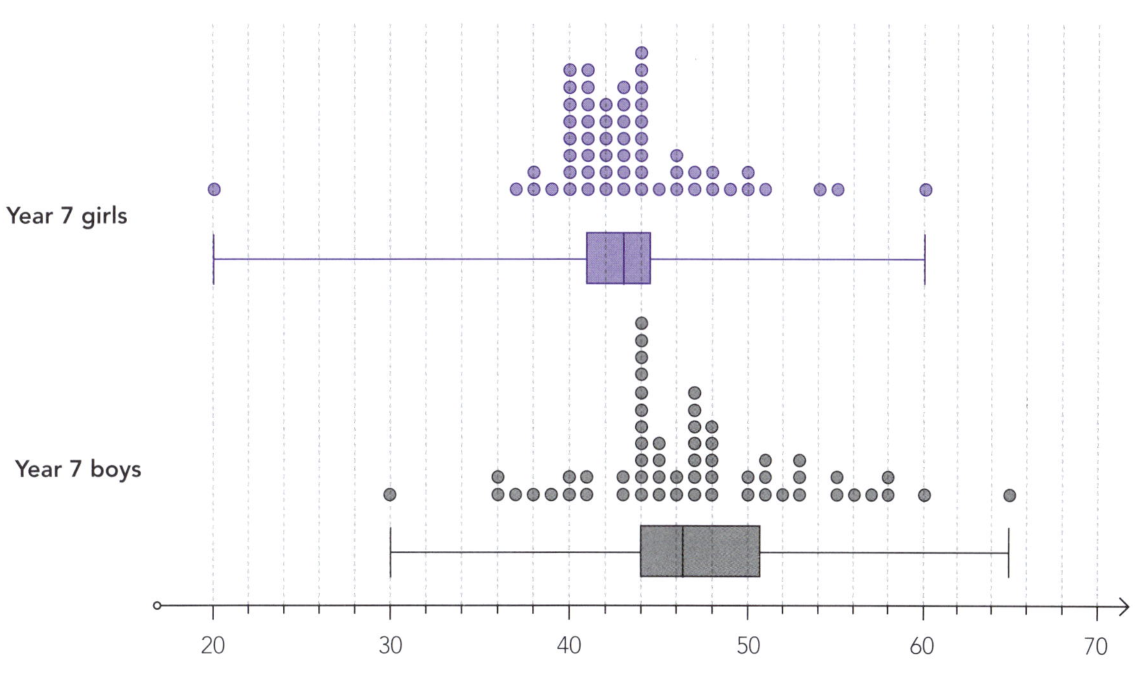

	Min.	LQ	Median	Mean	UQ	Max.	Sample size
Year 7 girls	20	41	43	42.95	44.5	60	58
Year 7 boys	30	44	46.5	46.84	50.5	65	58

Write a suitable question:

ISBN: 9780170370417

Centres

Shapes

Spreads

 ISBN: 9780170370417

Unusual features

Conclusion — make a call

Variability statement and reflections

Practice task two

The following shows the number of NCEA credits earned before the end of the first term in Year 11 by samples of 49 students at Paradise High School and 43 students at Heavenly High School.

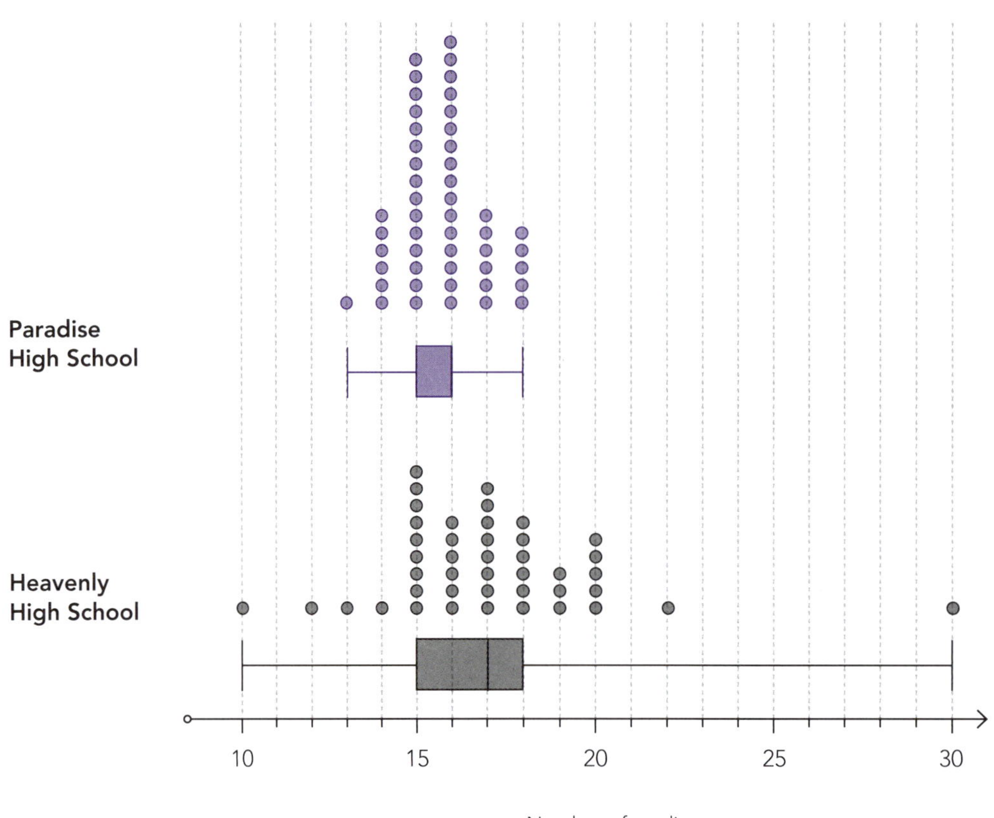

	Min.	LQ	Median	Mean	UQ	Max.	Sample size
PHS	13	15	16	15.71	16	18	49
HHS	10	15	17	17.05	18	30	43

Write a suitable question:

ISBN: 9780170370417

Centres

Shapes

Spreads

ISBN: 9780170370417

Unusual features

Conclusion — make a call

Variability statement and reflections

 ISBN: 9780170370417

Practice task three

The following shows heights (cm) for 57 boys who play table tennis and 38 boys who play volleyball at Paradise High School.

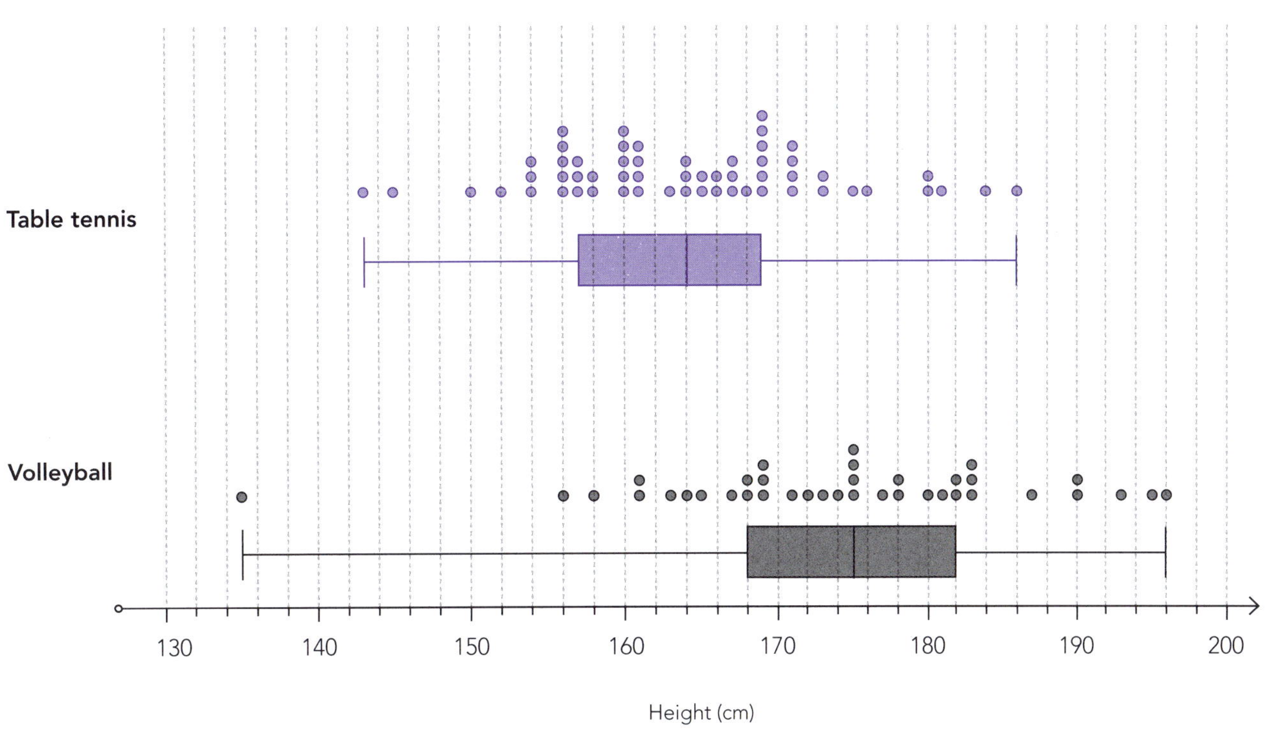

	Min.	LQ	Median	Mean	UQ	Max.	Sample size
Table tennis	143	157	164	164.1	169	186	57
Volleyball	135	168	175	174.2	182	196	38

ISBN: 9780170370417

Practice task four

Many New Zealanders own cats. Using the data below of a sample of cats from the town of Flockton, carry out a statistical investigation into some aspect of New Zealand cats.

- Pose an appropriate comparative question that can be answered from the data below.
- Either select a random sample or use all the data to calculate some appropriate statistics.
- Select and use appropriate display(s).
- Give summary statistics.
- Discuss and compare sample distributions.
- Communicate your findings in a conclusion; include an inference.

	Sex	Breed	Age (years)	Weight (kg)	Sleeps	Litter size	Collar
1	Female	Devon Rex	3.4	4.9	Inside	4	Yes
2	Male	Moggy	2.6	3.1	Outside	3	No
3	Male	Coon	0.7	2.1	Inside	3	Yes
4	Female	Siamese	1.2	2.4	Inside	2	Yes
5	Female	Burmese	10.2	4.9	Inside	5	No
6	Female	Moggy	5.6	3.7	Outside	6	No
7	Female	Persian	4.3	3.8	Inside	3	Yes
8	Male	Moggy	12.4	6.5	Outside	4	Yes
9	Female	Bengal	6.0	4.7	Outside	1	Yes
10	Male	Siamese	0.5	1.8	Outside	2	Yes
11	Male	Persian	4.8	4.9	Outside	7	No
12	Male	Moggy	7.2	5.3	Outside	4	No
13	Female	Burmese	9.1	5.9	Inside	7	Yes
14	Male	Birman	5.2	5.6	Inside	2	Yes
15	Female	Moggy	8.3	4.2	Inside	5	Yes
16	Male	Manx	2.6	3.7	Outside	3	No
17	Male	Moggy	4.8	5.1	Outside	7	No
18	Female	Moggy	1.0	2.2	Outside	2	No
19	Female	Burmese	6.9	3.8	Outside	4	Yes
20	Female	Siamese	11.3	4.9	Inside	5	Yes
21	Male	Moggy	9.7	5.8	Outside	2	No
22	Female	Burmese	2.3	4.0	Inside	5	Yes
23	Male	Moggy	4.3	6.3	Inside	3	Yes
24	Female	Bengal	2.1	3.5	Outside	4	No
25	Female	Persian	4.0	4.1	Outside	2	No
26	Male	Devon Rex	1.9	3.3	Inside	1	Yes
27	Female	Moggy	2.6	3.8	Inside	7	Yes
28	Male	Birman	5.2	4.9	Inside	5	No
29	Male	Siamese	0.3	1.0	Inside	4	No
30	Male	Moggy	2.8	2.6	Outside	2	Yes

ISBN: 9780170370417

	Sex	Breed	Age (years)	Weight (kg)	Sleeps	Litter size	Collar
31	Male	Moggy	7.2	4.7	Inside	4	No
32	Female	Burmese	3.6	4.1	Outside	4	No
33	Female	Burmese	6.1	3.7	Inside	6	No
34	Male	Manx	2.3	3.2	Outside	4	No
35	Male	Moggy	4.7	4.3	Inside	2	Yes
36	Female	Moggy	1.7	2.6	Outside	1	Yes
37	Male	Coon	2.4	4.6	Outside	3	Yes
38	Female	Moggy	5.3	4.3	Outside	4	No
39	Female	Siamese	5.1	3.5	Inside	2	Yes
40	Male	Moggy	9.3	5.1	Inside	3	Yes
41	Female	Siamese	4.6	3.6	Inside	3	No
42	Female	Burmese	6.1	4.2	Inside	5	No
43	Male	Moggy	7.8	5.1	Outside	2	Yes
44	Male	Persian	2.0	4.6	Inside	1	Yes
45	Male	Birman	3.1	4.2	Inside	4	Yes
46	Female	Moggy	1.5	3.2	Outside	3	No
47	Female	Burmese	5.4	3.7	Inside	2	No
48	Female	Burmese	2.2	3.0	Inside	5	Yes
49	Male	Moggy	5.7	4.3	Outside	4	Yes
50	Male	Burmese	3.1	4.1	Outside	6	Yes
51	Female	Birman	2.8	3.7	Inside	4	No
52	Female	Siamese	6.7	3.5	Inside	3	No
53	Male	Bengal	3.7	6.2	Inside	5	Yes
54	Female	Moggy	1.6	2.9	Outside	2	No
55	Male	Moggy	0.9	1.5	Inside	6	Yes
56	Male	Moggy	3.4	4.1	Inside	4	No
57	Female	Burmese	2.4	3.2	Outside	4	No
58	Male	Siamese	7.1	4.6	Outside	3	No
59	Female	Moggy	6.3	4.1	Inside	6	No
60	Female	Moggy	2.7	3.6	Inside	2	Yes
61	Male	Moggy	8.1	3.4	Outside	6	No
62	Female	Siamese	2.8	4.1	Inside	3	Yes
63	Male	Burmese	2.1	5.8	Inside	2	Yes
64	Male	Moggy	4.0	4.6	Outside	7	No
65	Male	Moggy	3.5	3.8	Outside	5	Yes
66	Female	Burmese	6.0	3.9	Inside	3	Yes
67	Male	Moggy	7.2	3.5	Outside	3	No
68	Female	Moggy	2.4	4.1	Outside	4	No
69	Male	Persian	1.0	2.0	Inside	2	Yes
70	Female	Moggy	0.6	1.4	Outside	5	No

ISBN: 9780170370417

Write a suitable question:

Select and highlight your sample.

Describe how you selected your sample:

Calculate your summary statistics and display your data.

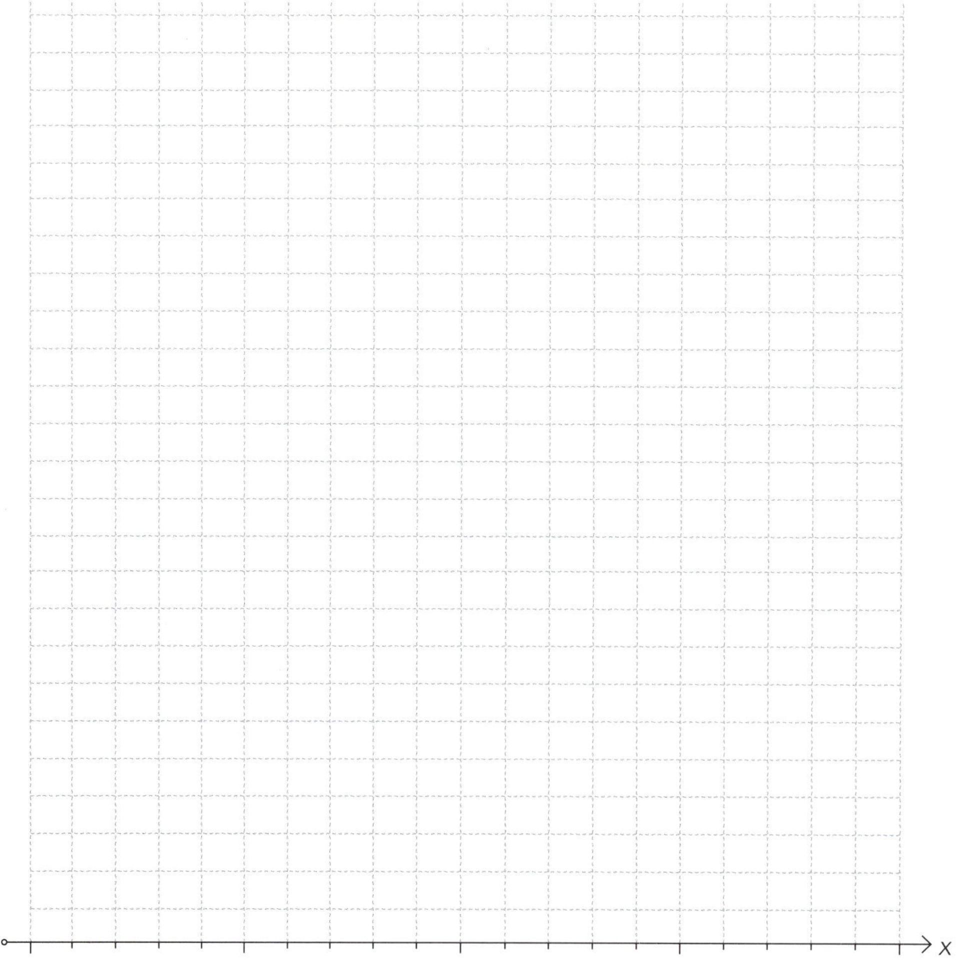

 ISBN: 9780170370417

Answers

Statistics (pp. 7–11)

Measures of centre (p. 7)

1 Mean = 2.6
Median = 2.5
Mode = 1
The best are the mean and the median because they are close to each other and there are no extremely big or small values.

2 Mean = 17
Median = 18.5
Mode = 20
The best is the median because it is near the centre of the data. The mode is much too high. The mean is too low because it has been dragged down by the 2, which is much smaller than any other value.

Measures of spread (pp. 8–11)

1 Minimum = 1
Lower quartile = 3
Median = 6
Upper quartile = 10
Maximum = 12
Range = 11
Inter-quartile range = 7

2 Minimum = 15
Lower quartile = 18
Median = 21
Upper quartile = 25
Maximum = 26
Range = 11
Inter-quartile range = 7

3 Minimum = 0
Lower quartile = 2
Median = 5
Upper quartile = 7
Maximum = 9
Range = 9
Inter-quartile range = 5

4 Minimum = 28
Lower quartile = 31
Median = 34.5
Upper quartile = 43
Maximum = 46
Range = 18
Inter-quartile range = 12

5 Minimum = 1.5
Lower quartile = 1.6
Median = 1.85
Upper quartile = 2.2
Maximum = 2.5
Range = 1
Inter-quartile range = 0.6

6 Minimum = 5
Lower quartile = 24
Median = 45
Upper quartile = 56
Maximum = 99
Range = 94
Inter-quartile range = 32

7 Minimum = 1
Lower quartile = 4
Median = 7
Upper quartile = 12
Maximum = 24
Range = 23
Inter-quartile range = 8

8 Minimum = 18
Lower quartile = 21
Median = 24
Upper quartile = 25.5
Maximum = 28
Range = 10
Inter-quartile range = 4.5

9 Minimum = 0
Lower quartile = 1.5
Median = 3
Upper quartile = 4.5
Maximum = 9
Range = 9
Inter-quartile range = 3

10 Minimum = 23
Lower quartile = 27
Median = 48
Upper quartile = 84.5
Maximum = 112
Range = 89
Inter-quartile range = 57.5

11 Minimum = 21
Lower quartile = 50.5
Median = 59
Upper quartile = 70

ISBN: 9780170370417

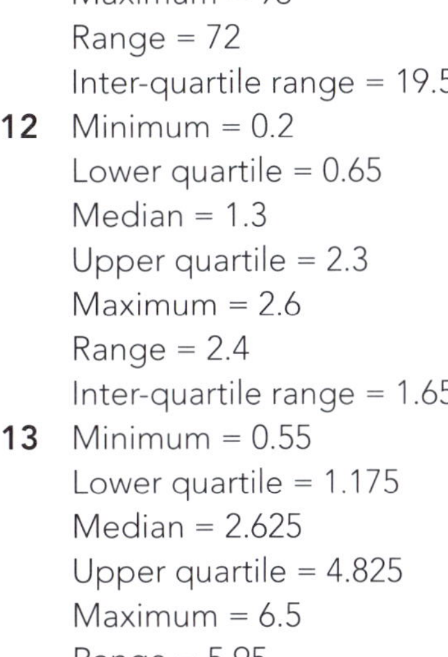

Maximum = 93
Range = 72
Inter-quartile range = 19.5

12 Minimum = 0.2
Lower quartile = 0.65
Median = 1.3
Upper quartile = 2.3
Maximum = 2.6
Range = 2.4
Inter-quartile range = 1.65

13 Minimum = 0.55
Lower quartile = 1.175
Median = 2.625
Upper quartile = 4.825
Maximum = 6.5
Range = 5.95
Inter-quartile range = 3.65

Display of data (pp. 12–20)

Dot plots (p. 12)

1

2

3

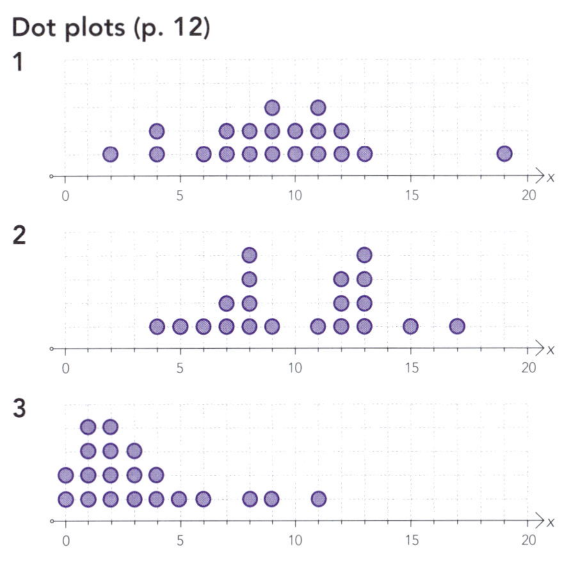

Box plots (pp. 13–14)

1

2

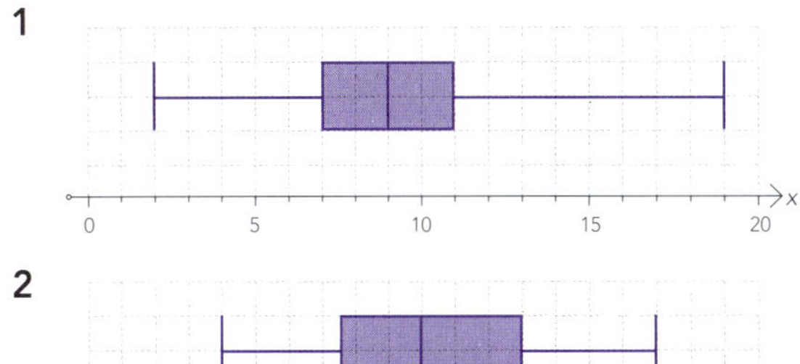

3

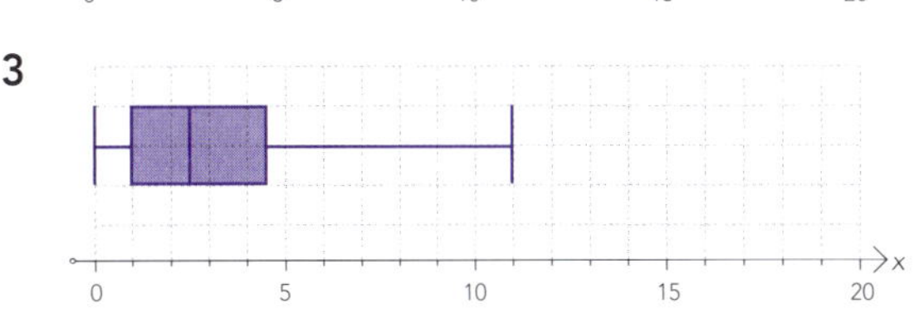

4

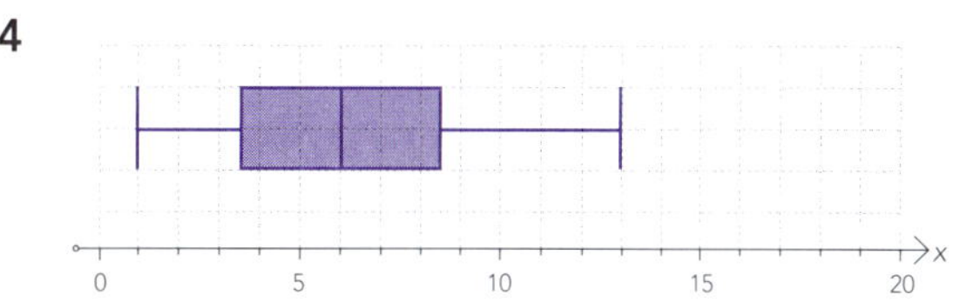

Putting it together (pp. 15–18)

1

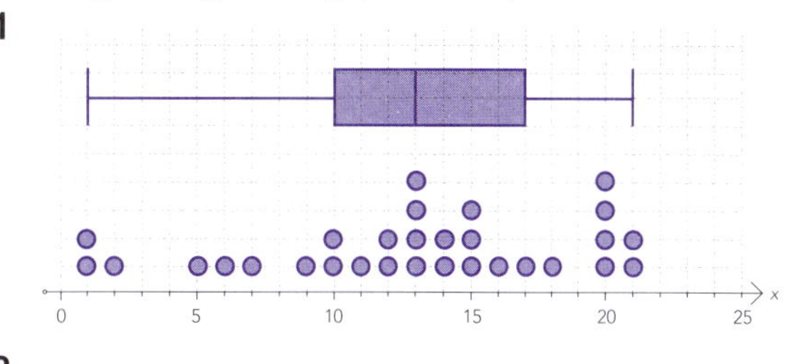

2

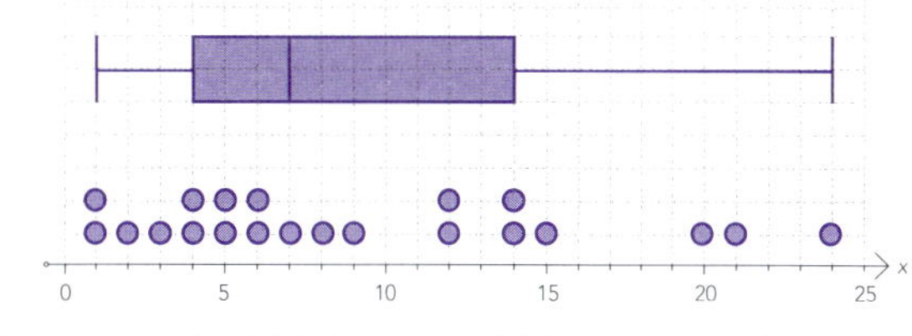

3

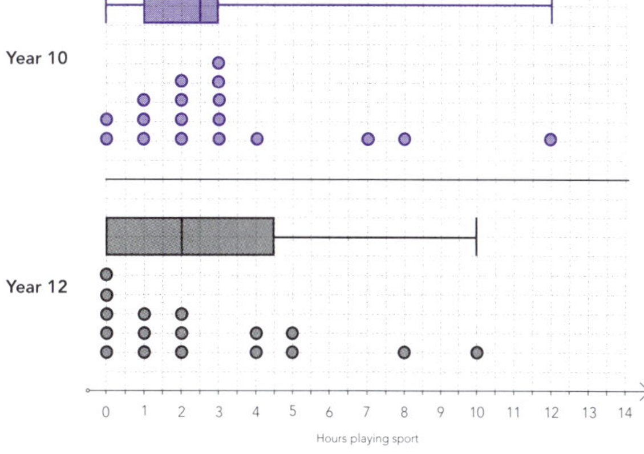

4

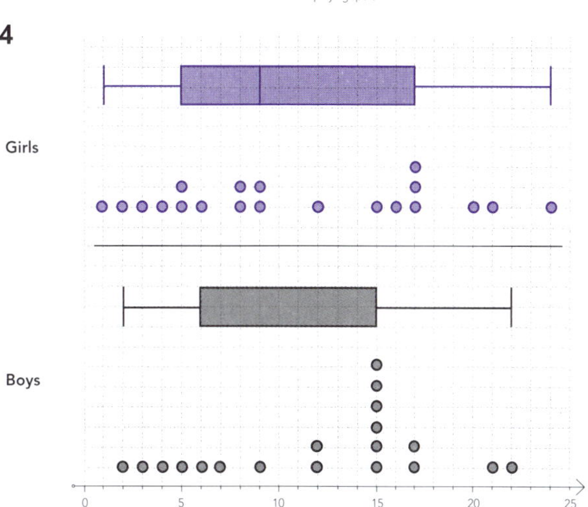

ISBN: 9780170370417

Combining dot plots, box plots and means (p. 19)

1 Mean is about the same as the median (**b**)
2 Mean is below the median (**c**)
3 Mean is below the median (**a**)

Box plot interpretation (p. 20)

1 Because the upper quartile is the same as the maximum value, in this case 9.
2 Because the minimum value is the same as the lower quartile, in this case 1.
3 Because the median is the same as the upper or lower quartile. We cannot tell which, but it could be 2 or 5.

Thinking about dot and box plots (p. 20)

1 Dot plots show the value of every data point. Therefore they show the shape of the distribution. For instance, they show whether a distribution is bimodal, and whether there are unusual points which lie a long way from the rest of the data. These cannot be seen on a box plot.
2 Box and whisker plots show the values of the upper and lower quartiles and the median. Therefore the position of the middle 50% of data can be seen easily. The width of the box (IQR) is also obvious, and this gives a measure of the spread.
3 Both show whether the data is symmetrical or skewed to one side. Both show the range.

Planning an investigation (pp. 21–31)

Writing questions (pp. 21–23)

	Yes/No	Why?	Make it better
Are 13-year-old boys taller than 13-year-old girls?	No	No population	Do 13-year-old boys tend to be taller than 13-year-old girls at Paradise High School?
What is the relationship between a student's arm span and height?	No	No population No direction	I would like to know if the arm spans tend to be greater than the heights of students at Paradise High School.
I would like to know how the heights of Year 11 boys compare with the heights of Year 11 girls at Paradise High School.	No	No direction	I would like to know if the heights of Year 11 boys are greater than the heights of Year 11 girls at Paradise High School.
I would like to know if the bag weight of Year 11 girls tends to be greater than the bag weight of Year 12 girls.	No	No population	I would like to know if the bag weight of Year 11 girls tends to be greater than the bag weight of Year 12 girls at Paradise High School.
What is the difference in hair of a Year 9 class at Paradise High School?	No	No variable specified — what about hair? Doesn't have two groups	Does the hair of Year 9 students tends to be longer than that of Year 10 students at Paradise High School?
Do boys take longer to get to school than girls?	No	No population	Do boys tend to take longer to get to school than girls at Paradise High School?
I would like to know the difference between how boys and girls at Paradise High School get to school.	No	How students get to school (bus, bike, walk, etc.) is not a numeric variable. No direction	I would like to know if students at Paradise High School who travel to school by bus take longer to get to school than students who bike.

ISBN: 9780170370417

Check with your teacher if your questions are not on this list:

- I would like to know if, in Flockton, cats that sleep outside tend to be heavier/older/have bigger litters than cats that sleep inside.
- I would like to know if, in Flockton, cats that wear a collar tend to be heavier/older/have bigger litters than cats that don't wear a collar.
- I would like to know if, in Flockton, male cats tend to be heavier/older than female cats.
- I would like to know if, in Flockton, moggies tend to be heavier/older/have bigger litters than pure bred cats.

Collecting data (pp. 24–25)

1 Just one class might not represent the whole of Year 10. It might be a top class, or a class that studies a particular group of subjects, e.g. technology. So this would be biased.

2 The students who walk past the senior common room are more likely to be senior students, so junior students may be under-represented in the sample, so it would be biased and not reflect the whole school.

3 Students who are late, enter by another gate or are absent will not have a chance of being selected, so the sample would be biased.

4 This should give an unbiased sample of school students because all are equally likely to be selected.

5 This surveys only those who use the canteen, so presumably like the food, and can afford to buy it. Those who don't use the canteen will not be surveyed. It will not get the views of those who buy from the canteen at times other than lunch time.

6 The survey will be self-selected. Only those who are interested enough, have the time and can be bothered will respond.

7 This might produce a random sample, but it might result in more or fewer students of particular nationalities being selected, e.g. the letter L could favour Korean and Chinese students.

8 Because the students are selected randomly, this should produce an unbiased sample. However, if there are significantly different numbers of students in each level, the levels with smaller numbers might be over-represented.

Sampling variation (pp. 29–31)

3 They should look different because every sample will contain a different selection of cats, so their statistics will vary.

4 Mason's sample was not random. Cats that sleep inside may tend to be heavier or lighter, or tend to have longer or shorter lifespans.

7 The samples of 10 should be more variable than the samples of 20.
You are more likely to get a group of unusually big or little values in a small sample than in a large sample. A large sample is more likely to have individuals from the whole range of values, and is more likely to reflect the population.

Sampling variation summary (p. 31)

1 Every sample taken from a population will be different.
If I repeated my sampling and analysis, I would get different results.
If I took a **bigger** sample:
- My results would be more reliable.
- The mean and median would probably be closer to the population mean and median.
- The IQR would probably be closer to the population IQR.

If I took a **smaller** sample:
- My results would be less reliable.
- The mean and median would probably be further from the population mean and median.
- The IQR would probably be further from the population IQR.

Describing and comparing features (pp. 32–55)

Centres

1 Comparing centres using statistics (pp. 32–33)

1 Boys: Median 19 Mean 18.69
Girls: Median 26 Mean $25.0\dot{6}$
Both the mean ($25.0\dot{6}$) and the median (26) times for girls are larger than the mean (18.69) and median (19) for boys. This means that girls tend to take longer to get ready for school in the morning than boys.

ISBN: 9780170370417

2 Boys: Median 158.5 Mean 160.4
Girls: Median 154.5 Mean 154.1
In these samples boys tend to be taller on average because their median is 4 cm higher than that for girls. This conclusion is also supported by the means: the mean height for boys is 6 cm more than the mean height for girls.

3 Boys: Median 16.5 Mean $17.\dot{6}$
Girls: Median 16 Mean 16.05
Because the medians are very similar, they suggest that on average boys and girls in these samples tend to take about the same times to get to school. However, the mean time taken for boys to get to school is about 1½ minutes longer than that for girls. This suggests that boys may tend to take slightly longer to get to school.

2 Comparing centres using dot plots (pp. 34–35)

1

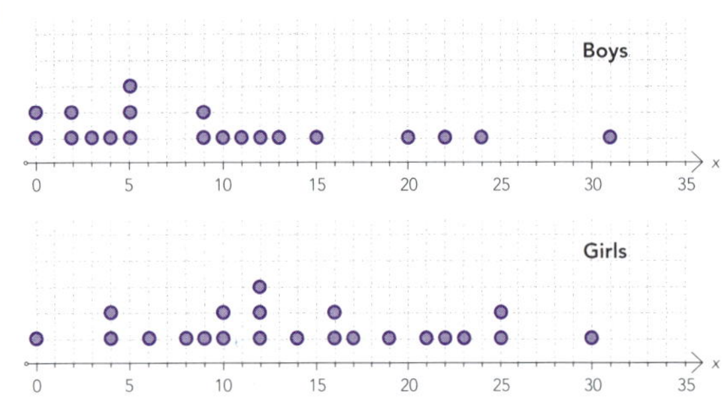

In these samples, the girls tend to spend longer doing their homework, because their dots are spread fairly evenly between 0 and 30, whereas most dots for the boys are between 0 and 15 minutes.

2

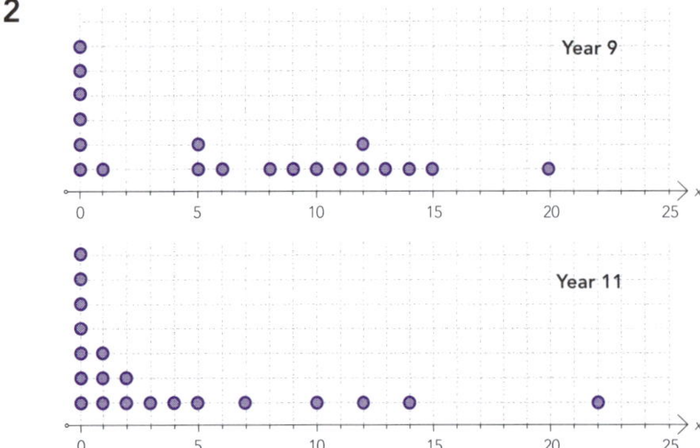

For these samples, the Year 9 students tend to be absent more frequently because most dots for Year 11 are clustered to the left, indicating absences of between 0 and 5 days. Most dots for Year 9 are between 5 and 15 days.

3

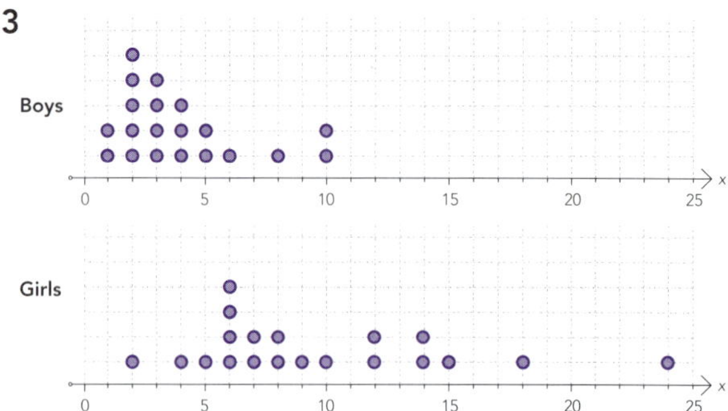

For these samples, the girls tend to own more shoes because their dots are mostly between 5 and 15, but the dots for the boys are strongly clustered to the left, with all between 1 and 10.

3 Comparing centres using box plots (pp. 36–38)

1

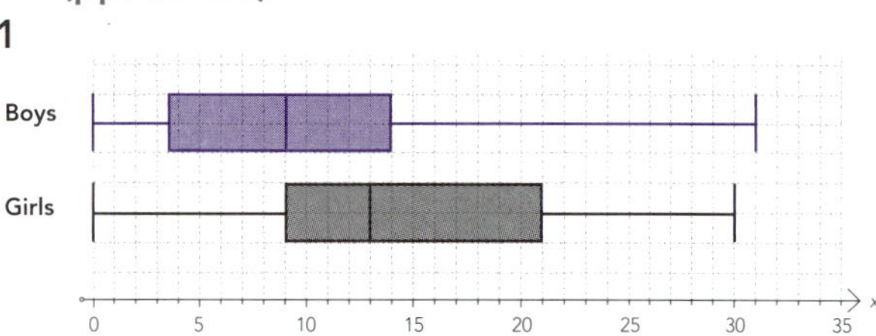

In these samples, girls tended to spend more time on average doing homework because the median for girls (13 minutes) is to the right of that for boys (9 minutes). Also the box for the girls' times is further to the right than that for boys.

2

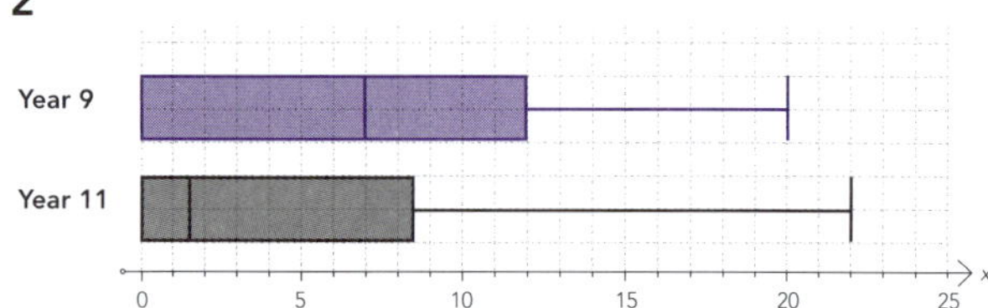

In these samples, Year 9 tend to have more days absent because their median (7) is to the right of that for Year 11 (1.5). Also the box for Year 9 extends much further to the right than that for Year 11.

3

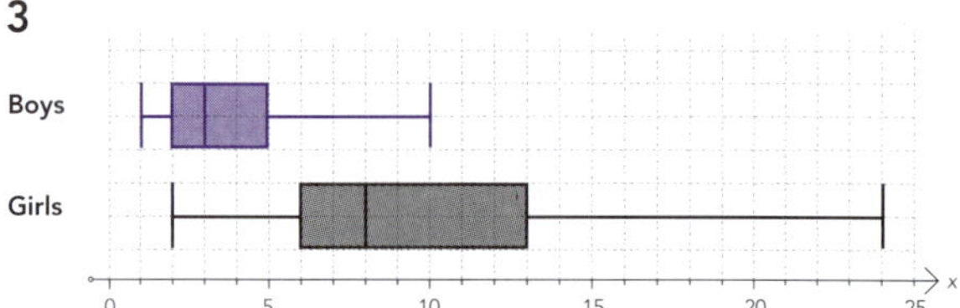

In these samples, the median number of pairs of shoes owned by girls (8) is to the right of that for the boys (3), so girls tend to own far more shoes than boys. This is supported by

ISBN: 9780170370417

the positions of the boxes: entire box for the girls is to the right of that for boys.

4

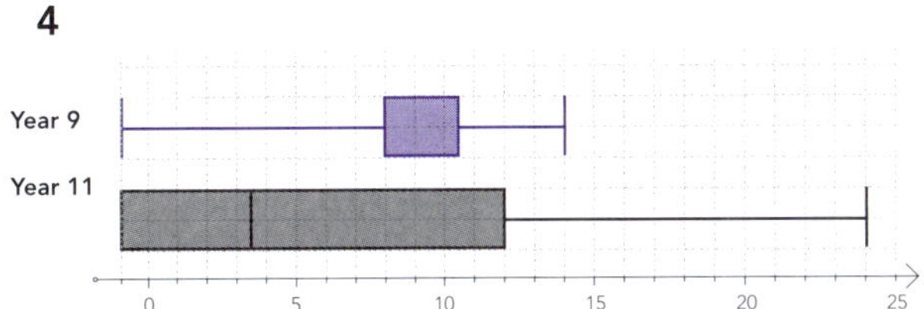

In these samples, the median time spent by Year 9 (8 minutes) is further to the right than that for Year 11 (4.5 minutes), so Year 9 tend to spend longer on their Science homework. The positions of the boxes support this because that most of the Year 9 box is further to the right than most of the Year 11 box.

5

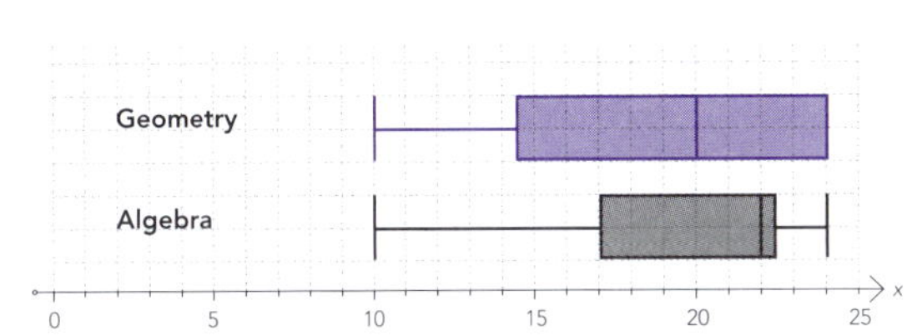

The median for Algebra (22) is further to the right than that for Geometry (20), so on average the class tended to get higher marks in Algebra.

Shape (pp. 39–42)

1 Describing shapes (pp. 39–40)

1. The data tends to be skewed to the left.
2. The data tends towards a normal distribution.
3. The data tends towards a uniform distribution, with clusters at the extremes.
4. The data has an irregular distribution.

2 Comparing shapes (pp. 41–42)

1. The girls tend towards an irregular distribution, but with a slight skew to the right. The boys tend towards a normal distribution. This means that the boys' marks tend to be clustered around the median, whereas the girls' marks are slightly clustered to the left.
2. The girls' marks tend towards a normal distribution, but the boys' marks are bimodal. This means that the girls' marks are clustered in the middle, but there are two clusters of boys' marks: around 7 and 15–16.
3. The girls' marks form a uniform distribution, and the boys' marks are irregularly distributed. This means that the girls' marks are evenly spread, whereas the boys' marks are irregularly spread.

Spreads (pp. 43–44)

1. The box for the girls is to the right of that for the boys. Both the IQR and the range are bigger for the girls. There is no overlap of the boxes. This means that in these samples, the girls did much better than the boys, although their marks were more spread out than the boys' marks.
2. The box for the girls is to the right of that for the boys. The ranges are similar, but the box for the girls is a little wider than that for the boys. The third quarter of the boys' marks overlaps with the second quarter of the girls' marks. This means that in these samples, the girls did better overall than the boys, but their marks were a little more spread.
3. The IQRs are close, but the range for the girls is a little bigger. The box for the boys lies within the box for the girls, and their positions are similar. This means that in these samples, the boys and girls got similar marks, although the girls' marks are slightly more spread.

Unusual features (pp. 45–46)

1. The value at 24 is an unusual point because it is so far from the rest of the data.
2. The two values at 2 form a small group which is some distance from the rest of the data.
3. There are no unusual points or groups. The data is clearly skewed to the left, so the point at 1 can be considered part of the skew.

Cleaning data (p. 46)

1. Remove 281 cm because it is not a feasible arm span for a Year 11 student.
2. 1.85 cm is not a feasible arm span for a Year 11 student, but 185 cm is, so assume the measurement was made in m and change 1.85 to 185.
3. 211 cm is a very wide arm span for a Year 11 student, but not impossible, so leave this in the data.
4. All the measurements are rounded to 0 dp except 189.5, so round that to 190 cm and leave it in the data.

Putting it together (pp. 47–55)

1 Centres

Both the mean and median female arm spans are bigger than those for boys. The dots for female arm spans tend to be further to the right than those for males. The box for females is slightly further to the right than that for boys. This means that, in these

ISBN: 9780170370417

samples, the arm spans for the girls tend to be bigger than those for the boys.

Shapes

Apart from the four smallest values, the girls' arm spans tend towards a normal distribution because most values are in the middle. The boys' arm spans tend towards a very wide normal distribution, so although there are more values in the middle, they are also very widely spread.

Spreads

The dots for the boys' arm spans are more spread than those of the girls. Also, both the IQR for boys (22 cm) and the range (82 cm) are bigger than those for the girls (14.5 and 52 cm respectively). This confirms that the arm spans for the boys are more widely spread than those of the girls.

The box for the girls' arm spans overlaps completely with the box for the boys' arm spans, but it is slightly further to the right. This means that, for the sample, the middle 50% of the girls arm spans lie within the middle 50% of the boys' arm spans.

Unusual features

There is one group of girls who have unusual values: four girls all have arm spans of 132 cm, which is at least 10 cm less than any of the other girls.

2 **Centres**

Both the median (25 cm) and the mean (24.82 cm) of the boys' right feet are greater than those for the girls (23 cm and 23.57 cm respectively). The dots for the boys tend to be further to the right than those for the girls, and the box for the boys is also further to the right. This means that, for these samples, boys tend to have longer right foot than girls.

Shapes

The shapes of both distributions tend towards the normal distribution. This means that for both boys and girls right feet, most values are in the centre. The values for the girls tend to be slightly skewed to the right, and for the boys, slightly skewed to the left.

Spreads

The IQRs for both distributions are 3 cm, so the spreads of boys' and girls' right feet are about the same. However, the range for the girls (18 cm) is slightly larger than that for the boys (14 cm). This is because there is one girl with a very short right foot (15 cm).

The box for the boys' feet lies further to the right than that for the girls. The third quarter of the female foot length overlaps with the second quarter of the boys' data. Both these facts show that boys in the sample tend to have longer right feet than girls in the sample.

Unusual features

As mentioned above, there is one girl with a very short right foot (15 cm). This point is unusual because her foot is at least 5 cm smaller than that of any other girl.

There are also two boys with unusually large right feet. Both have right feet that are at least 5 cm bigger than any other boys.

3 **Centres**

Both the median (54) and the mean (61.98) of the number of texts sent by girls are greater than those for the boys (42.5 and 47.48 respectively). The dots for the girls tend to be further to the right than those for the boys, and the box for the girls is also further to the right. This means that, for these samples, girls in Year 11 at Paradise High School tend to send more texts than boys.

Shapes

Apart from one girl who sent 352 texts, the shapes of the two distributions are very similar, and skewed to the right. The dots in both are mostly clustered towards the left, and the right tails of the distributions are more spread.

Spreads

The IQR for the girls (32.5) is bigger than that for boys (18). This means that the number of texts sent by girls in a week is more spread than that for boys. The range for the girls (324) is much larger than that for the boys (72). This is because there is one girl who sent huge numbers of texts (352).

The box for the number of texts sent by girls lies further to the right than that for the boys. The second quarter of the number of texts sent by girls overlaps with the most of the box of the boys' data. Both these facts also show that girls in the sample tend to send more texts in a week than boys in the sample.

Unusual features

The only unusual feature is the girl who sent 352 texts in a week. I think this is unusual because it is well over double the number of the nearest value (about 125 texts).

4 **Centres**

Both the median (44) and the mean (44.79) of the number of the shoulder widths of Year 12

 ISBN: 9780170370417

students are greater than those for the Year 9 (40.5 and 41.04 respectively). The dots for the Year 12 students tend to be further to the right than those for Year 9, and the box for the Year 12 students is also further to the right. This means that, for these samples, Year 12 students at Paradise High School tend have wider shoulders than Year 9 students.

Shapes

Both distributions tend towards the normal distribution because the data is more tightly grouped between the medians, and more spread out outside them. However, because the numbers in the samples are relatively small, the shape is rather irregular.

Spreads

The IQR for the Year 12 students (7.5 cm) is slightly bigger than that for the Year 9 students (6.5 cm). This means that the spread of shoulder widths is slightly greater for the Year 12 students.

The box for the Year 12 students is slightly further to the right than that for the Year 9 students. This supports the conclusion that for these samples, Year 12 students have wider shoulders than Year 9 students.

The third quarter for the Year 9 students overlaps with most of the middle 50% for the year 12 students.

Unusual features

There are no unusual features in this data.

Making the call about the population (pp. 56–64)

Method 1 (pp. 56–58)

1 We can conclude that lizards in population A are likely, on average, to be longer than those in population B because, although the middle 50% of each sample overlap, the median of A lies outside the box of B.

2 We cannot conclude that lizards in population A tend to be longer than those in population B, because both of the medians lie within the middle 50% of the other sample, and the boxes overlap.

3 We can conclude that lizards in population B will tend to be longer on average than those in population A because the boxes do not overlap and both medians lie outside the other box.

4 We can conclude that lizards in population A are likely, on average, to be longer than those in population B because, although the middle 50% of each sample overlap, both medians lie outside the box of the other group.

5 We can conclude that lizards in population B are likely, on average, to be longer than those in population A because, although the middle 50% of each sample overlap, the median of A lies outside the box of B.

6 We can conclude that lizards in population A will tend to be longer on average than those in population B because the boxes do not overlap and both medians lie outside the other box.

7 We can conclude that lizards in population B are likely, on average, to be longer than those in population A because, although the middle 50% of each sample overlap, the median of A lies outside the box of B.

8 We cannot conclude that lizards in population B tend to be longer than those in population B, because both of the medians lie within the middle 50% of the other sample, and the boxes overlap.

Method 2 (pp. 59–61)

1

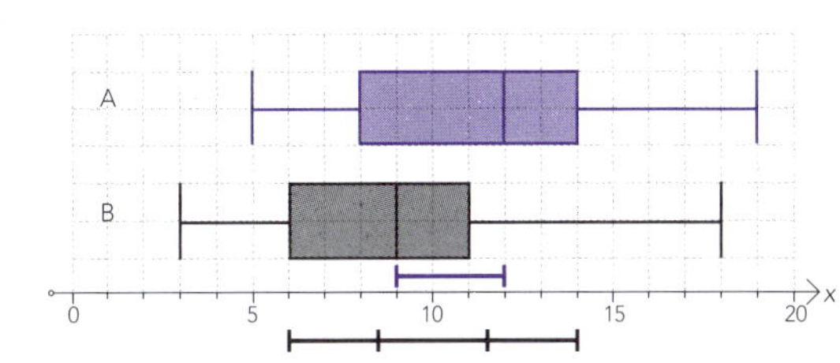

The DBM is bigger than ⅓ of the OVS so we can conclude that lizards in population A **will** tend to be longer than those in population B. This is a different conclusion because by Method 1, we only concluded that a difference was likely.

2

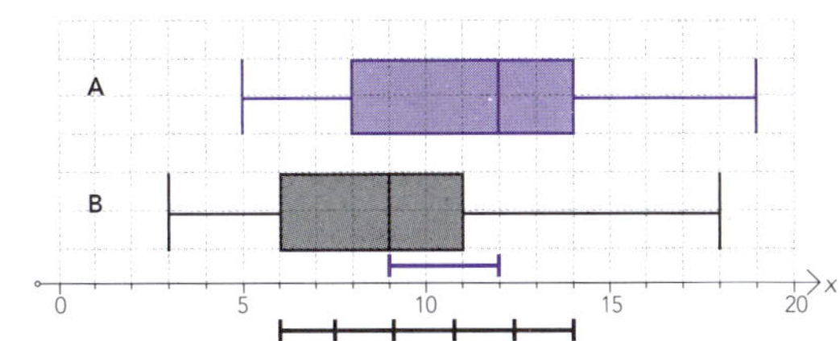

The DBM is much bigger than ⅕ of the OVS so we can conclude that lizards in population A **will** tend to be longer than those in population B.

This is a different conclusion because by Method 1, we only concluded that a difference was likely.

3

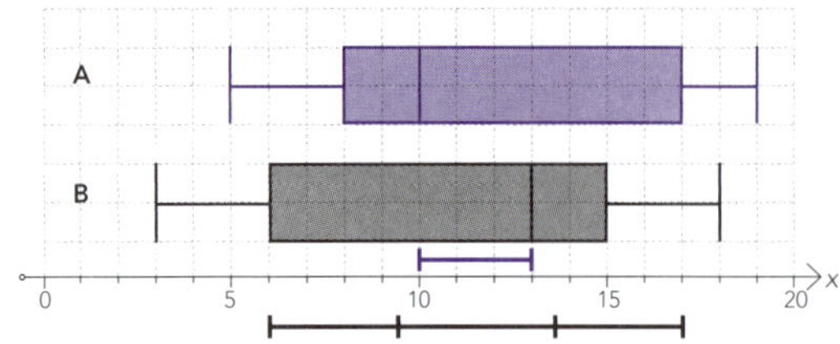

The DBM is not bigger than ⅓ of the OVS so we cannot conclude that lizards in population B will tend to be longer than those in population A.
This is the same conclusion as for Method 1.

4

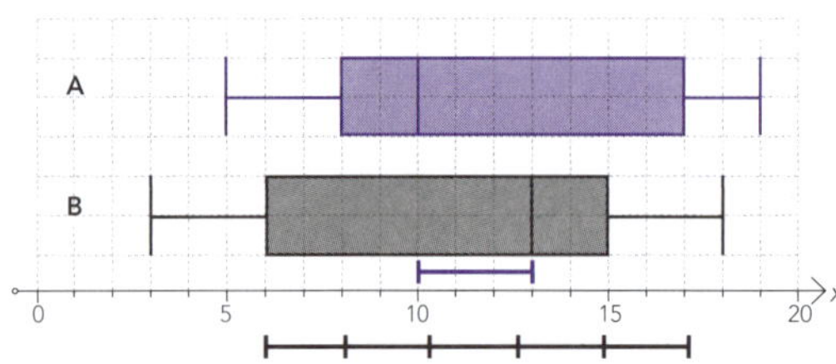

The DBM is bigger than ⅕ of the OVS so we can conclude that lizards in population B will tend to be longer than those in population A.
This is not the same conclusion as for Method 1.

5

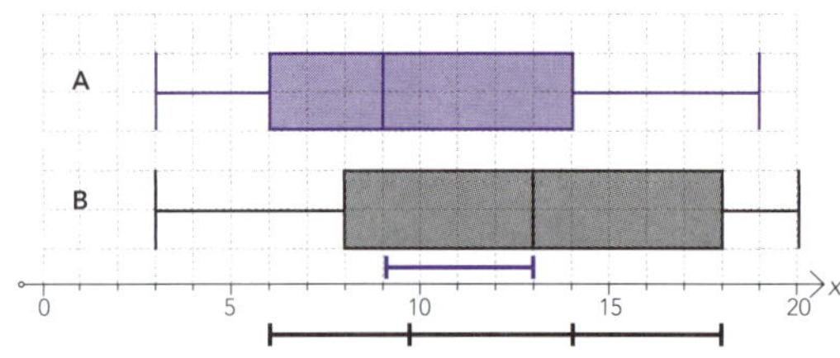

The DBM is not bigger than ⅓ of the OVS (they are the same) so we cannot conclude that lizards in population B will tend to be longer than those in population A.
This is the same conclusion as for Method 1.

6

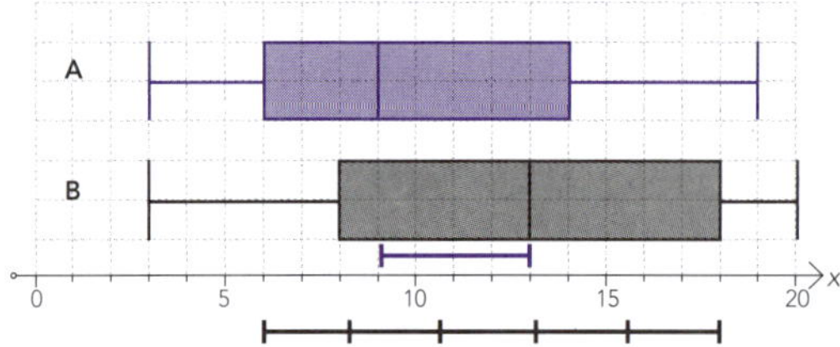

The DBM is bigger than ⅕ of the OVS so we can conclude that lizards in population B will tend to be longer than those in population A.
This is not the same conclusion as for Method 1.

Questions and conclusions about the population (pp. 62–64)

1 **Question**

Do the boys in Year 13 tend to be taller than the girls at Paradise High School?

Conclusion

Method 1

Both medians lie outside the other box, so we can conclude that in the population boys in Year 13 at Paradise High School will tend to be taller, on average, than the girls.

Method 2

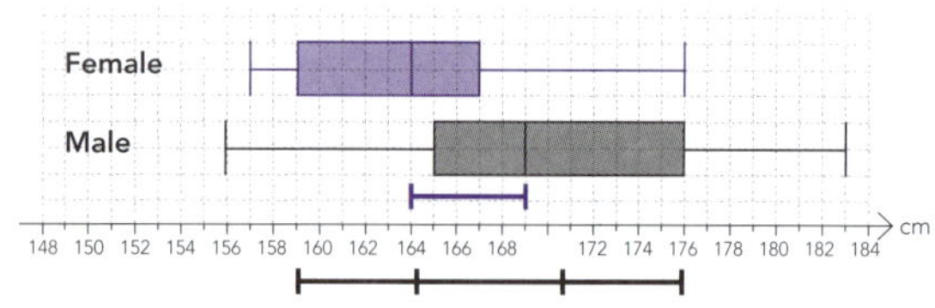

The DBM is not bigger than ⅓ of the OVS so we cannot conclude that in the population, boys at Paradise High School will tend to be taller than the girls.
Note: The methods give different results.

2 **Question**

Are the hourly wages earned by boys in Year 13 at Paradise High School higher than those of girls at Paradise High School?

Conclusion

Method 1

Both medians lie within the other box, so we cannot conclude that in the population boys in Year 13 at Paradise High School tend to earn more than girls.

Method 2

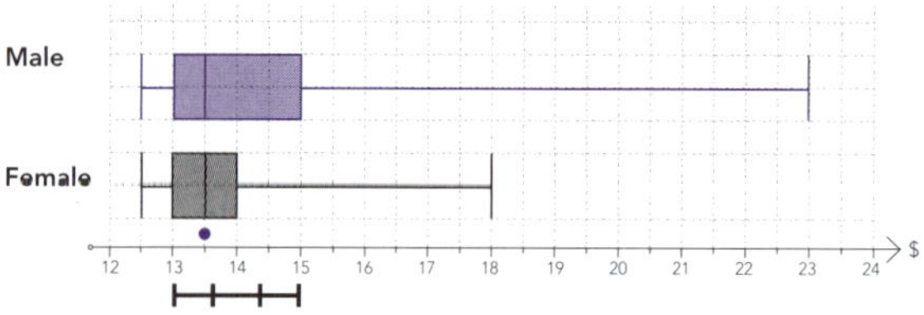

The DBM = 0, so we come to the same conclusion as for Method 1.

3 **Question**

Do girls in Year 13 at Paradise High School send more texts in a week than boys?

Conclusion

Method 1

We can conclude that in the population, girls in Year 13 at Paradise High School are likely, on average, to send more texts in a week than boys, because the median for the girls lies outside the box for the boys.

  ISBN: 9780170370417

Method 2

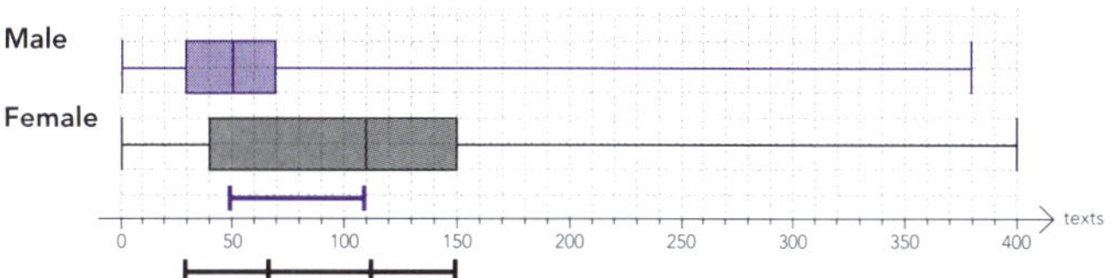

The DBM is bigger than ⅓ of the OVS so we can conclude that in the population, girls in Year 13 at Paradise High School do tend to send more texts in a week than the boys in Year 13 at Paradise High School.

Writing about data (pp. 66–70)

1 B

2 A, B and G (7)
C and F (8)
D and H (6)

3 E and G
H and G

4 B with G
E with G
H with G

5 B with G
E with G
H with G

6 B with G
E with G

7 C and H

8 B with D
H with D

9 Doesn't make sense because you can't compare the location of a point (median) with an interval (IQR) which has width but no location.

10 False. The conclusion is the wrong way around. Individuals in D would tend to be bigger than those in E.

11 False. The IQRs are the same (7). However, the statement would be correct if the first sentence was 'The IQR of A is **further to the right** than the IQR of B.'

12 True

13 False. It is skewed to the left.

14 True

15

Common mistakes	What is wrong with them?
The medians overlap.	Medians can't overlap as they are point values; they are either the same or different.
There is an unusual point so I will ignore/remove it.	Just because it is unusual doesn't mean it is wrong. Think about whether it is possible for the given context.
The boys' median is bigger than the girls' so there is a difference in the population.	This isn't necessarily the case. If the boys' median is larger than the girls' UQ, then there is likely to be a difference in the population.
If I did another sample, I would get the same results.	The statistics are likely to be different but the conclusion should be the same if the sample size is adequate.
I have 30 in each group, so my conclusion is right.	While 30 is the minimum number, it doesn't necessarily mean that your conclusion will reflect the population. The larger the sample, the more confident we can be.
The UQ of the girls is bigger than the boys. This means the shape is irregular.	The use of 'This means' in this case is wrong. Don't link things that aren't related.
The Year 9 inter-quartile range is to the right of the Year 11 median.	The interquartile range is a measure of spread. It doesn't describe the location of the middle 50% of the data.
There is a tail to the right, therefore the shape is a left skew.	Wrong way round: if the tail is to the right, then the graph is a right skew.

Practice tasks (pp. 71–83)

Practice task one

A suitable question:

Do New Zealand Year 7 boys tend to have longer little fingers than Year 7 girls?

Centres

Both the mean (46.84 mm) and the median (46.5 mm) for the boys' little fingers are greater than those for the girls (42.95 mm and 43 mm). This means that for these samples, the boys tended to have longer little fingers than the girls.

Shapes

Apart from one girl who had a very short little finger (20 mm), the distribution for the girls is relatively normal, although **slightly** skewed to the right. The dots to the right of the UQ are more spread out than the dots to the left of the LQ.

The boys' distribution is irregular, although fairly symmetrical. Although it is irregular, the dots between the quartiles are more tightly grouped than those outside the quartiles.

Spreads

The IQR for the boys (7 mm) is bigger than that for the girls (4 mm), and had it not been for the 20 mm finger of one of the girls, the range for the boys would have been bigger as well (35 mm vs about 23 mm). This means that in these samples, the boys' little finger lengths are more spread than the girls' finger lengths.

The box for the boys lies further to the right than that for the girls, although the third quarter for the girls overlaps with the second quarter for the boys.

This supports the earlier conclusion, that for these samples, the boys tended to have longer little fingers than the girls.

Unusual features

As mentioned earlier, the only unusual feature **in these samples** is the girl with a 20 mm little finger. I think this is unusual because the next biggest girl's finger is about 37 mm long, which is almost double.

This doesn't seem to be a very likely length for a girl's little finger, but it is possible that she is a dwarf, or that she has had part of her little finger chopped off.

Conclusion — make a call

Both medians lie outside the other box, but the boxes overlap, so we can conclude that in the population of New Zealand Year 7 students, boys are likely to have longer little fingers, on average, than the girls. This does not surprise me as boys tend to be a little taller than girls.

OR

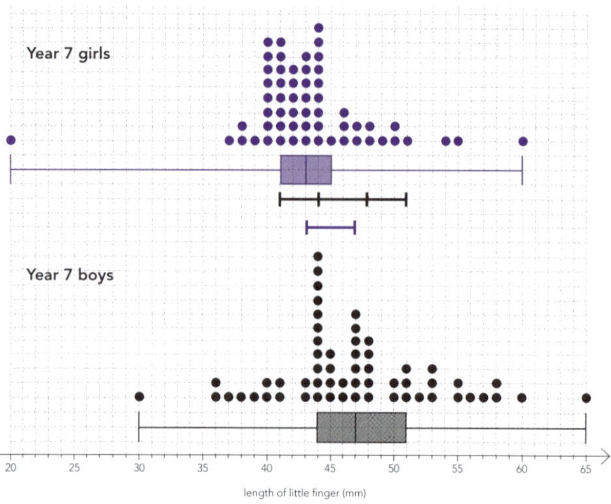

The DBM is not bigger than ⅓ of the OVS so we cannot conclude that in the population of New Zealand Year 7 students, boys will tend to have longer little fingers than the girls. I am surprised by this result because in general, boys tend to be a little taller than girls.

Variability statement and reflections

Only 58 Year 7 girls and 58 Year 7 boys were in the samples, so if we took a different sample, we would not expect exactly the same results, although, if the samples were randomly selected, then the results would probably be similar. Bigger samples would probably give us more reliable results.

Other possible points for discussion or further investigation:

- Would you expect similar results if a different finger was used?
- Why might the girls' data be skewed to the right?
- Are the differences similar in older students?
- Why did the two methods for making the call give different results?

Practice task two

A suitable question:

Do the Year 11 students as Heavenly High School tend to earn more NCEA credits during the first term than those at Paradise High School?

Centres

Both the mean (17.05) and the median (17) number of credits earned by Year 11 students at Heavenly High School are greater than those for students at Paradise High School (15.71 and 16). This means that for these samples, students at

ISBN: 9780170370417

Heavenly High School tended to earn more credits during the first term than the students at Paradise High School.

Shapes

Both distributions of credits in these samples tend towards a normal distribution. The data for Paradise High School is grouped more closely. The number of credits earned by Year 11 students at Heavenly High School is slightly less regular.

Spreads

The IQR for Paradise High School is 1, and for Heavenly High School it is 3. This means that for these samples, there is a much bigger spread in the number of credits earned at Heavenly High School. This is supported by the huge difference in the ranges (20 at Heavenly High School and only 5 at Paradise High School). However, this is partly caused by one person at Heavenly High School who got 30 credits. Without this value, the range for Heavenly High School would be about 12, which is still much bigger than that for Paradise High School.

The box for Heavenly High School lies further to the right than that for Paradise High School. The entire box for Paradise High School lies inside the second quarter of Heavenly High School's data. This supports the earlier conclusion that for these samples, students at Heavenly High School tended to earn more credits during the first term than the students at Paradise High School.

Unusual features

There is one unusual point **in these samples:** one person at Heavenly High School got 30 credits. I think this is unusual because this person got 8 more than anybody else. It seem unlikely that one person should get so many more credits than anybody else. One possible explanation is that they were repeating Year 11, and had earned some credits the previous year.

Conclusion — make a call

One median lies outside the other box, but the boxes overlap, so we can conclude that in the populations, Year 11 students at Heavenly High School are likely to earn more NCEA credits in term 1 than those at Paradise High School. This result surprises me, because I would expect the number of credits earned to be similar in different schools.

OR

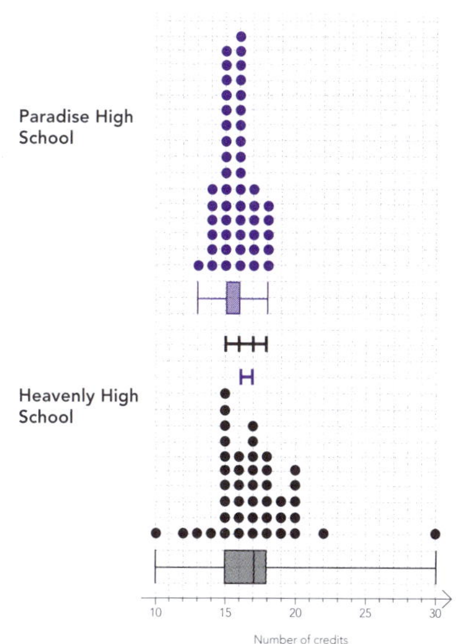

The DBM is not bigger than ⅓ of the OVS so we cannot conclude that in the populations, Year 11 students at Heavenly High School are likely to earn more NCEA credits in term 1 than those at Paradise High School. This result does not surprise me, because I would expect the number of credits earned to be similar in different schools.

Because of the big differences in the IQRs and ranges, and the difference in the medians, it looks as though Paradise High School are offered fewer credits during the first term, but most people pass them. At Heavenly High School, it looks as though they offer a lot more credits in term 1, but more people do not achieve all of them.

Variability statement and reflections

There was data for only 49 Year 11 students at Paradise High School and 43 from Heavenly High School. If we took a different sample we would not expect to get exactly the same results. Our results would probably have been more reliable if we had larger samples. It would depend on how the samples were selected. If the 49 was data from just two classes, then the results might be quite different if a different method of selection was used.

If the samples were randomly selected, then the results would probably be similar. Also bigger samples would **probably** give us more reliable results.

Other possible points for discussion or further investigation:

- Were there differences the number of credits earned by students at the two schools during the other three terms?
- How did the two schools compare at the end of the year?

ISBN: 9780170370417

- Is the pattern the same at levels 2 and 3?
- Why did the two methods for making the call give different results?

Practice task three

A suitable question

Do boys who play **table tennis** at Paradise High School tend to be taller than those who play volleyball?

Centres

Both the mean (174.2 cm) and the median (175 cm) heights for the boys who play volleyball are greater than those of the table tennis players (164.1 cm and 164 cm) at Paradise High School. This means that, for these samples, boys who play volleyball tend to be taller than those who play table tennis.

Shapes

The shape of the distribution of heights of table tennis players is irregular. However, the heights between the quartiles are more tightly grouped than those outside the quartiles, as found in a normal distribution. The shape of the distribution of heights of table volleyball players is irregular.

Spreads

The IQR for volleyball players (14 cm) is slightly larger than that for table tennis players (12 cm). This means that for these samples, the heights of volleyball players are a little more spread than those of the table tennis players. This is supported by the ranges: 61 cm for volleyball players, and 43 cm for table tennis players.

The box for volleyball players is further to the right than that for table tennis players. This, along with the differences in the medians and means, supports the conclusion that, for these samples, volleyball players tend to be taller than table tennis players at Paradise High School.

The boxes overlap a little: the upper part of the third quarter of the heights of the table tennis players overlaps with the lower part of the second quarter of the volleyball players' heights.

Unusual features

There is one boy who plays volleyball who is much shorter than the rest (135 cm). He is about 20 cm shorter than any other player. This is possible because he may be the libero for the team. The libero is a specialist defender who needs to be able to get under the ball, even when it is near the floor. For this reason he doesn't need to be tall.

There were no unusual features in the table tennis data.

Conclusion — make a call

Because both medians lie outside each other's boxes and the boxes overlap, it is likely that those in the population of volleyball players tend to be taller than those in the population of table tennis players at Paradise High School. This is what I would have expected because, apart from the libero, it is definitely an advantage to be tall for volley ball because the net is quite high. While a taller person usually has greater reach, that is not quite so important in table tennis.

OR

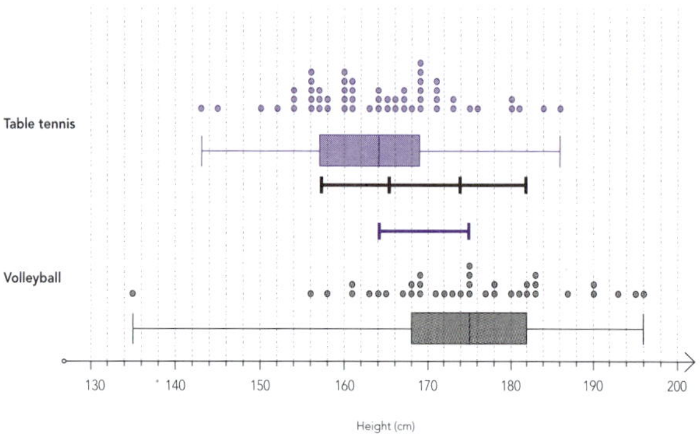

The DBM is not bigger than ⅓ of the OVS so we cannot conclude that those in the population of volleyball players tend to be taller than those in the population of table tennis players at Paradise High School. This is not what I would have expected because, apart from the libero, it is definitely an advantage to be tall for volley ball because the net is quite high. While a taller person usually has greater reach, that is not quite so important in table tennis.

Variability statement and reflections

Only 57 table tennis players and 38 volleyball players were sampled. If we took another sample we would not expect to get the same result. However, provided those in the samples were randomly chosen, the results would probably be similar, especially for the table tennis players because the sample was bigger. Our results would have been more reliable if we had larger samples from both groups.

Other possible points for discussion or further investigation:

- What ages were the players?
- How were they selected for the sample?
- Were they recreational or competitive players?
- Why did the two methods for making the call give different results?

  ISBN: 9780170370417